Free Will and Consciousness in the Multiverse

Christian D. Schade

Free Will and Consciousness in the Multiverse

Physics, Philosophy, and Quantum Decision Making

Springer

Christian D. Schade
School of Business and Economics
Humboldt University of Berlin
Berlin, Germany

ISBN 978-3-030-03582-2 ISBN 978-3-030-03583-9 (eBook)
https://doi.org/10.1007/978-3-030-03583-9

Library of Congress Control Number: 2018962787

© Springer Nature Switzerland AG 2018
This work is subject to copyright. All rights are reserved by the Publisher, whether the whole or part of the material is concerned, specifically the rights of translation, reprinting, reuse of illustrations, recitation, broadcasting, reproduction on microfilms or in any other physical way, and transmission or information storage and retrieval, electronic adaptation, computer software, or by similar or dissimilar methodology now known or hereafter developed.
The use of general descriptive names, registered names, trademarks, service marks, etc. in this publication does not imply, even in the absence of a specific statement, that such names are exempt from the relevant protective laws and regulations and therefore free for general use.
The publisher, the authors and the editors are safe to assume that the advice and information in this book are believed to be true and accurate at the date of publication. Neither the publisher nor the authors or the editors give a warranty, express or implied, with respect to the material contained herein or for any errors or omissions that may have been made. The publisher remains neutral with regard to jurisdictional claims in published maps and institutional affiliations.

Cover image: © Oliver Klimek/Fotolia (Adobe Stock)

This Springer imprint is published by the registered company Springer Nature Switzerland AG
The registered company address is: Gewerbestrasse 11, 6330 Cham, Switzerland

Acknowledgements

Let me start with my gratitude to Angela Lahee, my book editor at Springer Physics, for her competent support, including the constructive and inspiring discussions we had. The book in its current form would not have been realized without her. I am also grateful to the referees and their careful reading of and commenting on my manuscript; they clearly contributed to an improvement of it.

I also acknowledge the royalty-free policy of the *Journal of Cognition and Neuroethics*, that printed an article containing an early version of some of my thoughts: Schade, Christian D. 2015, "Collecting Evidence for the Permanent Coexistence of Parallel Realities: An Interdisciplinary Approach." *Journal of Cognition and Neuroethics* 3: 327–362. Although many of my ideas and concepts have been refined (or simply altered) since then, some premises (and some graphs) still draw from this paper.

I am especially thankful to Tanja Schade-Strohm. Tanja recently published a German language, spiritual and philosophical book on how to become the best version of yourself ("Werde zur besten Version deines Selbst: Aus der Verstimmung in die Bestimmung", Bielefeld: TAO 2017). This book is based on a multiverse idea of our reality, and her book inspired some ideas in my book (visible via a couple of references). She also personally contributed with various discussions about important aspects of this book as well as ideas, testified via a number of acknowledgement footnotes. I am also grateful for her long-term emotional support.

I finally thank Ute Ottenbreit who helped me with getting some of the literature underlying this book as well as some formatting, proofreading, etc., of the manuscript.

Contents

List of Figures

List of Tables

List of Boxes

Chapter 1
Introduction: Developing a Multiverse View of Decision Making and Consciousness

Scientific Revolutions and Narcissistic Wounds of Mankind

Our (western) worldview has been challenged a few times in history by scientific revolutions. The three most radical changes have been identified by Sigmund Freud in the move from a geocentric to a heliocentric perspective, in the understanding that mankind developed from the animal kingdom (instead of being purposely created), and in the detection of the power of the unconscious for our decisions (Freud 1917). Sigmund Freud called those three radical changes narcissistic wounds (*narzistische Kränkungen*). Others have identified more narcissistic wounds inflicted by science on mankind (Vollmer 1999), most importantly, for the topic of this book, *neurobiological reductionism*, or, as Vollmer calls it, the dissolution of the dualism of body and soul. Closely related, the reductionist' worldview in general has been identified as one of four *injuries* by Kauffman (2010, 8).[1]

This book is concerned with the multiverse interpretation of quantum mechanics. Quantum mechanics is about a hundred years old and the existence of a plethora of worlds is the most logical interpretation of quantum mechanics as this book is going to show in Chaps. 2 and 3 (see for this perspective also, e.g., Deutsch 1997; Lockwood 1991; Mensky 2010; Wallace 2012a, b, c; Zeh 2012, 2013, 2016a). One might hence ask why the multiverse is not already part of mankind's general worldview. It is not even the dominant interpretation of quantum mechanics among physicists but ranks second[2] after the practically successful but theoretically

[1]The other three injuries are the division between the humanities and science, the negative view on spirituality and the split between the world of fact and the world of values (Kauffman 2010, 7–8). The emergentist' view that Kauffman (2010) then develops throughout his book as a means to heal those wounds will, however, not be considered within the current monograph. I rather aim at healing those wounds via the development and application of the concept of the clustered-minds multiverse (to be introduced in Chap. 4).

[2]I have to admit that such a ranking is hard to be made. Some physicists have conducted polls at conferences, but this hardly qualifies as 'hard data.' My subjective view, however, is that many physicists would agree with this 'ranking.'

© Springer Nature Switzerland AG 2018

C. D. Schade, *Free Will and Consciousness in the Multiverse*,
https://doi.org/10.1007/978-3-030-03583-9_1

implausible singular-world ‘standard interpretation’ (comprising the ‘reduction postulate’ à la John von Neumann and Paul Adrien Maurice Dirac, often subsumed under the umbrella of the ‘Copenhagen interpretation’[3]). Also, many physicists have hoped to be able to *circumvent* the interpretation problem of quantum mechanics somehow via the application of the *decoherence principle*—an unsuccessful attempt (see Chap. 2 of this book).[4]

According to Squires (1994), however, there are reasons for this situation that somehow relate to the above idea by Freud (1917) and others: “Could it be that the reluctance of many physicists to (…) recognize the existence of a multitude of universes, is of similar nature to the earlier reluctance to accept that our world is not the unique centre of all things?” (133). Let me add: If many physicists are reluctant to accept the idea of a multitude of universes, or better: realities,[5] what do we expect from people with a background in other sciences or non-scientists? Let me furthermore add: It is probably fair to assume that ‘of similar nature’ is implicitly associated by Squires with people being afraid of the next narcissistic wound: What if our reality is not the only one? What does it mean to talk of parallel realities?

But do we really have to be afraid of finally accepting the multiverse interpretation of quantum mechanics? I do not think so. And part of what this book intends to do is to put the reader in a position to decide herself whether or not to accept it; and if he decides to accept it, what to do with it in terms of weltanschauung, decision making etc. Sure enough, the multiverse perspective has to be digested, somehow, it radically challenges our current view of reality, and it is thus not ‘plug and play.’ Adopting the version of the multiverse view developed in Chap. 4 of this book: the clustered-minds multiverse, the existence of parallel realities might at least not feel ‘strange.’

I would like to argue that there is, anyway, *no good alternative* to accepting some multiverse interpretation of quantum mechanics or another in the long run if we accept quantum mechanics as an *explanatory model* at all—and not only as some epistemic, empirically predictive and ‘useful’ model in the spirit of a ‘shut-up-and-calculate’[6] mentality. And many indeed argue that it is highly adequate—and furthermore exciting—to take quantum mechanics seriously as an explanatory model, e.g., Lockwood (1991):

---

[3]According to Zeh (2013), this ‘umbrella view’ of the term Copenhagen interpretation is historically misleading. He also points to the fact that the actual proponents of the Copenhagen interpretation (Bohr, Heisenberg) maintained a quasi-classical view on quantum phenomena (particles), whereas von Neumann was using the wave function as consequently as possible. The ‘umbrella view’ has nevertheless gained some popularity, probably because all sub-interpretations handled under this umbrella are singular-world views without postulating third variables.

[4]Indeed, decoherence does not solve the measurement problem. However, it is quite helpful in crystallizing it (Zeh 2011).

[5]According to my own and many other authors’ view, there is still only one quantum multiverse, allowing, however, for many parallel realities (see Chaps. 2, 3 and 4).

[6]It is not quite clear to whom this ‘dictum’ might be attributed, many think that it was first expressed by Paul Dirac.

> (…) the world is quantum-mechanical through and through; and (…) the classical picture of reality is, even at the macroscopic level, deeply inadequate. It is true that the bulk of macroeconomic phenomena admits, to a high degree of approximation, of being analyzed in classical terms. But quantum mechanics is not to be regarded as just another scientific theory. To the extent that it is correct, it demands a complete revolution in our way of looking at the world, more profound than was required by any previous scientific breakthrough: this is what makes it so exciting philosophically. (178)

Healing the 'Neurobiological Wound' and Overarching Materialism

The idea of parallel realities, of a multiverse, makes many people feel uneasy when they are confronted with it for the first time. But as this book is going to demonstrate, the multiverse interpretation of quantum mechanics—at least in the version to be proposed here—is coming along with positive aspects for how we see and live our lives since it is actually *healing* some of the above-mentioned narcissistic wounds: e.g., the neurobiological reduction wound and the overarching materialism, perhaps as the consequence of the sum of all of the above (and other) wounds.[7] (Note that whenever I talk about the multiverse, it is the quantum multiverse, not any other that has been suggested; see, for different types of multiverses, Tegmark 2004.)

And this is important. According to the famous Libet (e.g., 1985) experiments and more recent neuroscience results, the existence of an actual *free will* (over and above, perhaps, some last-second vetoing power by consciousness in motor control; see Chaps. 3 and 6 for more details) is denied in contemporary science, and some (such as Dennett 2003a) would even argue that *consciousness* is an illusion—turning Indian idealism, arguing that the outside world is an illusion or *Maya* whereas consciousness is the only 'real thing' (see Chap. 5), on its head. More generally, most of the scientific insights in the last five hundred years (just reconsider the above "*Kränkungen*") did not only 'dethrone' mankind but indeed led to a more and more materialistic (or reductionist'; see Kauffman 2010, 1–18) worldview. The world as well as the people living in it lost more and more of their 'soul' and perhaps part of the *meaning* mankind used to attach to life.[8]

Quantum mechanics, in the multiverse version developed in the current book, is able to bring back part of the world's non-materialistic aspects. And it potentially helps relieve people of their 'mental bondage.' Indeed, it shows that much of the limitations that individuals perceive are mental in nature.

---

[7]This perspective has also been inspired by Goswami (2015). In his popular science book on "Quantum Economics," Goswami argues that quantum mechanics might be able to 'heal' mankind from overarching materialism, especially in the context of economic behavior.

[8]"(…) the real world we live in is a world of fact without values." (Kauffman 2010, 8).

This book will propose a novel version of the multiverse interpretation: the *clustered-minds multiverse*, where an actual free will exists in the form of the existence of *alternative possibilities* and an influence of the individual on *what will be experienced to what extent*; consciousness is at the core of everything. Within this book, I will derive that we are not just 'victims of circumstances.' Therefore, an actual responsibility for our actions appears to be—at least partially—justified (see Chap. 7). Moreover, this book will demonstrate that most consistent with the clustered-minds multiverse is a philosophy-of-mind conception I am going to call *dualistic idealism*, a position where the mental (consciousness) and the material are different realms, but where the realm of consciousness plays the dominant role and where the physical is only a 'playing field' provided for our consciousness (see Chap. 5). The perspective offered is, anyway, freeing our mind. It removes the feeling, resulting from all of the narcissistic wounds, of being part of a huge clockwork with no importance and no possibility of changing our life, whatsoever.

Decision Making as the Core Paradigm

There are probably many fields to look at and to demonstrate the changes that the multiverse perspective suggests for our worldview. This book will look at the quantum multiverse mainly through the lens of decision making. It will not only show that our choices can freely be made, but it will more generally look at the way we make decisions in the multiverse. At the center of the decision sciences are, normally, a general decision-making framework, the notions of probability and utility, also strategic choices and all their applications in economics and other social interaction. This book will show that the perspective on choices will indeed be changed when adopting a multiverse view.

E.g., the notion of probability has to be reevaluated, conceptually, since the Schrödinger equation is deterministic and there simply is *no randomness* existing, physically. Since this clearly does not imply that we know with certainty what is going to happen next, and the Born rule of quantum mechanics—used to provide *probabilities* for measurement outcomes[9]—is often seen as the most accurate random prediction of all existing scientific theory, some clarifying thoughts are needed within this book. The interesting question is what one is supposed to do with the notion of probability when making decisions in the multiverse—business as usual, despite the conceptual complexities? The puzzling nature of probabilities within quantum mechanics was already part of what made Schrödinger alert and seems to have 'converted' him into a 'multiverser' within the last decade of his life (the following quote is based on a lecture historically preceding Everett's relative

[9]As will be demonstrated in Chap. 4, the best way of looking at those probabilities is as a prediction of relative frequencies of measurement outcomes along one decoherent history (one reality).

state theory crafting the fundament of the multiverse; in fact, Everett acknowledged to be explicitly building up upon Schrödinger's thoughts):

> Nearly every result [a quantum theorist][10] pronounces is about the probability of this or that (...) happening – with usually a great many alternatives. The idea that they be not alternatives but *all* really happen simultaneously seems lunatic to him, just *impossible*. He thinks that if the laws of nature took *this* form for, let me say, a quarter of an hour, we should find our surroundings rapidly turning into a quagmire, or sort of a featureless jelly or plasma, all contours becoming blurred, we ourselves probably becoming jelly fish. It is strange that he should believe this. For I understand he grants that unobserved nature does behave that way – namely according to the wave equation. (Schrödinger 1995 [1952], 19)

For many cognitive psychologists or economists working in the field of the decision sciences, the notion of *utility* (or value) *maximization* is major building block side-by-side with probability, at least within so-called *normative* models (see Chap. 8). As will be shown, the multiverse perspective requires a new decision-making framework, and this in turn has an effect on the appropriate notion of utility. However, the book will not be able to advise a fully revised (or even conclusive) picture but rather critically reflects upon different concepts of utility and their advantages and disadvantages in general as well as with respect to their usage within a multiverse framework. Additionally, it is quite challenging in this regard that consciousness is willing to make all kinds of experiences, not only the nice ones. Therefore, I will have to be very careful with interpreting what 'maximal utility' is.

The strategic and economic aspects of decision making are also quite 'tricky,' from the perspective of the multiverse. The reason is that one is looking at the joint result of many choices on a joint or an individual outcome of a game or the 'result' of a market, and that a quantum-based theory is lacking on how such choices might interact. Game theory might be helpful, here, as is the case in a singular universe, but assumes a certain type of rationality and makes standard assumptions on utility. It might be seen, however, as a starting point of the theoretical development required here. This is an exciting field for future research albeit a complex one requiring the effort of highly interdisciplinary researchers.

In a Nutshell: The Three Ingredients of Free Will

One of the most interesting aspects of decisions is whether there are actually any, i.e., whether they can freely be made. As I am going to demonstrate in some detail in this book, free will is possible on the basis of quantum mechanics, albeit the type of free will is different from what 'folk wisdom' might expect. Free will requires three ingredients:

[10] Insertion in the Schrödinger quote by Lockwood (1996), 165.

- The first ingredient is the *existence of alternative possibilities*. This implies selecting—or rather newly proposing—a certain form of the *multiverse interpretation* of quantum mechanics. Not all variants of the multiverse interpretation are equally free-will friendly. But an important implication of all of them is the existence of a plethora of *versions of yourself*,[11] versions of yours you are not aware of. Slightly simplifying the 'story' that I would have to tell later in this book, let me suppose that I am driving through a big city and decide to make two subsequent right turns. Then there is a version of mine that already took a left in the first decision and is now facing a different situation and different choices than I am, i.e., living in a parallel reality. Another version decided to take a right in the first decision, but took a left in the second etc. This is no play with thoughts but reality, when we follow the multiverse interpretation of quantum mechanics and a requirement for free will.[12]
- The second ingredient is *parallel times*. I am going to argue that *objectively* there is *no flow of time* from the past to the present to the future as we would normally suppose but that the flow of time is rather a *subjective*, a perceptual phenomenon. This thought is somewhat consistent with the *B-theory of time* in philosophy, often equated with *eternalism* (Gale 1966; McTaggart 1908; for a more recent exposition, partially based on modern physics: Barbour 1999). There is a specific reason for dealing with parallel times in the context of free will: the interdisciplinary perspective on the multiverse proposed in Chap. 3. A cornerstone of accepting the multiverse as the appropriate interpretation of quantum mechanics or completing the multiverse 'proof' proposed in Chap. 3—comprising the phenomenon of free will (see Fig. 3.1)—is physiological anticipation, the fact that our bodies are able to anticipate future events. This empirically robust phenomenon can *only* be explained if times are parallel, if different points in time are special cases of parallel realities, as will be shown. Moreover, some other proponents of the multiverse (not all!) share the parallel-time perspective: e.g., Deutsch (1997) and Mensky (2010). Theoretical reasons are *special relativity* (e.g., Minkowski 1952 [1908]; Petkov 2005) and the related (static, eternal) block-universe view (e.g., Silberstein et al. 2018) as well as *quantum gravity* (e.g., DeWitt 1967), related to the approach by Deutsch (1997).

---

[11]The usage of the term 'versions' will be preferred in this book over and above the usage of terms such as replicas, copies etc. used by other authors. The term 'versions' has, e.g., been used in the theoretical literature on quantum mechanics by Zeh (2013) as well as in Schade-Strohm (2017).

[12]The reader who has watched (and may remember) the movie "Run Lola Run" ("Lola rennt," with Franka Potente as Lola and produced by Tom Tykwer, Germany 1998) may feel reminded of the quick sketches of alternative lives based on slightly altered choices, provided within a few seconds, at several points throughout this film as well as the different outcomes based on different short 'histories,' explicitly considered within this film via totally altered decision sequences. In fact, this film gets surprisingly close to the multiverse perspective proposed in this book.

- The third ingredient is *dualistic idealism* (see Chap. 5). Whereas physical reality (i.e., the multiverse) is the *boundary condition* for the experiences we are able to make, there is no fixed coupling of a certain *degree* of consciousness with all possible realities 'out there.' Consciousness is allowed to decide how much emphasis to put on which reality. This concept will turn out to be better accessible to the reader after Chaps. 2, 3, 4 and 5.

This Book's Contribution to the Literature

The discussion as to how quantum mechanics should be interpreted has partially left the realm of physics and entered the realm of philosophy already a while ago. This statement unambiguously includes publications on the multiverse interpretation. The most important philosophical contributions so far are Albert and Loewer's (1988) claim that the 'multiverse interpretation needs interpretation' and the proposals they make, Lockwood's (1991) analysis of the mind-body problem as well as his many-minds theory (Lockwood 1996), both based on the multiverse, Barrett's (1999) thoughtful (albeit skeptical) monograph exploring some of the multiverse's theoretical consequences, as well as the so-called Oxford realist' interpretation(s) of the multiverse (e.g., Wallace 2012a, b, c; Saunders et al. 2012). Many contributions concerning itself with conceptual questions around the multiverse interpretation have been written by physicists, and they turn out to be closely related to the discussion in philosophy. These are, e.g., the monographs by Squires (1994), containing a long discussion on how to interpret quantum mechanics with a strong emphasis on the multiverse interpretation, Deutsch (1997, 2012b) and Mensky (2010).[13]

In the multiverse, a plethora of possibilities of making experiences exists. In the last section, this was the first ingredient listed as a requirement for the existence of free will. And indeed, in the development of his *Extended Everett Concept* (EEC), Mensky (2010) briefly deals with the question of the existence of free will and answers it in the affirmative. Important thoughts are added in Mensky (2013). However, Mensky's treatment of the free will problem is based on conceptually and ontologically problematic premises and cannot be adopted within this book as will be shown in Chap. 4 (see especially Box 4.1).

Only recently, David Deutsch (2012a) has stated that time is overdue to leave the discussion as to *whether* the multiverse interpretation is appropriate; but to enter the stage where its consequences be explored in various domains and thus its fruitfulness checked. I would like to argue that David Deutsch is right in asking for applications and explorations of the multiverse perspective. But as already discussed above, not everyone might be ready to accept the multiverse interpretation

[13]The decision as to who I am classifying as a physicist or a philosopher is a bit arbitrary, I suppose.

and would be willing to explore its consequences without being presented further evidence first. It thus appears that a monograph is overdue that collects such further evidence for the multiverse interpretation of quantum mechanics in a systematic way, that also links the multiverse interpretation *in detail* to the problem of free will, that furthermore offers a theory of the action of consciousness that might be judged as ontologically satisfying, and that, finally, explores various consequences of the multiverse view based on those theoretical developments. These issues are among the main contributions of this book.

Drawing from multidisciplinary sources (mainly from physics, philosophy, neurosciences, psychology, decision sciences and economics), this book will, however, limit its empirical applications (i.e., suggestions for experimental designs in future research etc.) mostly to physics and the social sciences (mainly psychology and economics). Its theoretical contributions go beyond that and are also relevant, e.g., to the discussion in philosophy and the neurosciences.

Overview of the Chapters

The following book chapters are now briefly introduced to allow for an overview of what will be addressed as well as an insight into the structure of the book. Note that within several chapters, this book will contain so-called 'boxes.' Boxes are used to draw the attention of the reader to important theoretical (sometimes controversial) issues or definitions within a condensed format.

Chapters 2, 3 and 4 are setting the stage for the multiverse interpretation suggested in this book: the clustered-minds multiverse, and constitute Part I of the book. Chapters 2 and 3 present *evidence* for the permanent coexistence of parallel realities. Specifically, Chap. 2 deals with the measurement problem of quantum mechanics and the resulting interpretation problem; it also concerns itself with the principle of decoherence and suggests a new perspective on it. It finally shows that, according to physical theory, the multiverse interpretation of quantum mechanics is the most *parsimonious* of the proposed interpretations (even though slightly less parsimonious as claimed by authors belonging to the Oxford interpretation). The chapter aims at being comprehensible for a large number of readers, even though I have to admit that despite leaving out any mathematical development, the degree of abstraction and complexity is unavoidably high at times. The chapter contains two boxes. Box 2.1 deals with the question whether quantum correlations (or decoherence) are to be interpreted top-down or bottom-up (Box 2.1), i.e., whether those correlations might start within consciousness. Box 2.2 discusses the question whether the multiverse interpretation is able to solve the locality problem of quantum mechanics. The chapter ends with presenting and discussing some objections others have raised against the multiverse interpretation of quantum mechanics as well support that has been crafted in favor of it.

Chapter 3 then takes an interdisciplinary perspective and aims at re-interpreting the famous Libet results showing that physiological processes are *preceding* conscious decisions and that are normally taken to imply an *impossibility of free will* (except, perhaps, some last-second veto possibility by consciousness in motor control). Closely related, it will also have a look at experimental findings on *anticipatory physiological responses* that might be seen as inconsistent with (classical) physics. Chapter 3 will contain a detailed discussion of different concepts of time in physics. This requires dealing with the block-universe view (based on special relativity) as well as quantum gravity (based on general relativity). The chapter will then suggest that different times are parallel (or, frankly, that time does not exist, depending on the applied perspective) and that this might solve the Libet puzzle and might offer a framework for understanding anticipatory physiological responses. However, the joint framework for anticipatory physiological responses and the existence of an actual free will will turn out to be the multiverse so that a singular-universe blockworld does not suffice in this regard. Chapter 3 additionally applies a teleological perspective with respect to the meaning, the *sense* of qualia, which might in principle be identified in the 'production' of free will. Chapter 3 is intended to be an interdisciplinary 'proof' of the multiverse, based on and extending Schade (2015). The chapter also contains two boxes. Box 3.1 addresses some objections against the notion of parallel times that may arise from the second law of thermodynamics and offers a new view of the problem. Box 3.2 deals with some criticism that others have raised with respect to seeing consciousness at the core of the measurement problem. Their objections will be countered.

Chapter 4 specifies the interpretation of the multiverse proposed in this book. There is no such thing as 'the' multiverse interpretation of quantum mechanics. Indeed, many scholars have pointed out that Everett's original 'many-worlds' interpretation needs interpretation. Existing versions exhibit different problems. A principle problem of all of them materializes in connection with the probability rule of quantum mechanics: the *Born* (1926) *rule*; a problem that my proposal cannot solve either, but where I am proposing a pragmatic handling. Most of them also have strange ontological consequences, most of them are not free-will friendly, most of them do not offer a satisfactory solution for the so-called *preferred-basis problem*, etc. Chapter 4 explains the problems that have been dealt with in the literature and then offers an interpretation of the multiverse without severe ontological problems that is also free-will friendly: the *clustered-minds multiverse*. As will be pointed out, the clustered-minds multiverse is a version of the multiverse interpretation where individuals' consciousness (depending on the preference for different macro-realities) has an impact on the degree of consciousness allocated to different realities (i.e., to different versions of the individual). The chapter will again be written in a way that makes the theoretical subtleties accessible also to readers that are no specialists in quantum mechanics. The chapter introduces the allegory of a torch light that helps visualizing the workings of consciousness in the multiverse according to the newly proposed interpretation; but it will also qualify and modify this simplifying picture. The chapter also contains a box. Box 4.1 deals with the question whether Menksy's (2005, 2010) proposal of free will via an individual's influence on subjective probabilities is theoretically acceptable.

The novel multiverse interpretation proposed in Chap. 4, the clustered-minds multiverse, has consequences for various discussions in philosophy. Those consequences will be explored—for some selected questions—in Chaps. 5, 6 and 7 (constituting Part II of this book). One of the old and new controversies in philosophy circles around the mind-body problem, being at the core of Chap. 5. Philosophers have taken distinctly different routes to tackle this problem. The clustered-minds multiverse suggests a dualistic view where consciousness (i.e., the mind) and the physical (i.e., the wave function) should be described as different realms. However, *consciousness* is seen as the starting point of entanglement (according to Chap. 2), it has an influence on what realities will be perceived to what extent and it makes the resulting experiences (i.e., qualia); thus, consciousness is *dominant*. Still, the physical offers the 'playing field' for consciousness, so that a radical position such as Indian idealism, albeit close to the concept proposed here, turns out not to be adequate. Thus, *dualistic idealism* is the most appropriate term for the mind-body-concept of the clustered-minds multiverse. Whilst developing this perspective, this chapter addresses other substantial problems that have been analyzed in the philosophy of mind. E.g., the idea of consciousness being supervenient on the physical will be rejected. The chapter also discusses and answers the question whether consciousness has an impact on the physical. The answer depends on whether a closed-system or an open-system perspective is taken on the universe or what will be defined as 'physical.' This issue is addressed in Box 5.1, located almost at the beginning of the chapter and one of two boxes in this chapter. In Box 5.2, the question is answered whether (weak) psychophysical parallelism would be an appropriate label for the consequences of the clustered-minds multiverse for the mind-body problem, too.

A plethora of possibilities of making experiences exist, and consciousness decides on how much emphasis to put on which realities in the form of the degree of consciousness allocated. Chapter 6 therefore argues that consciousness is indeed executing free will, albeit in a special form. Traditionally, there have been several quite different philosophical approaches dealing with free will, and the issue is still far from being settled. Simplifying, most theories can be classified according to three aspects: (a) whether they believe in determinism or not; and if they do, whether they (b) believe that free will is existing or not existing and (c) responsibility is justifiable or not justifiable under this condition; incompatibilists would normally deny (b) and (c), compatibilists would tend to agree with (b) and (c) (e.g., Nichols and Knobe 2007, and the literature referenced there). Chapter 6 builds upon Chaps. 2–5 and argues that under the conditions of the clustered-minds multiverse, some of those 'disputes' disappear: Whereas the world (as a total) is deterministic (indeed, the Schrödinger equation is!), people have an impact on how much consciousness will reside in which reality. This has consequences for the traditional classifications into libertarians, compatibilists, incompatibilists etc. Chapter 6 contains three more boxes. Box 6.1 discusses whether quantum brain biology offers an alternative possibility to 'save free will,' as has been claimed by Hameroff (2012). Box 6.2 discusses whether special forms of free will may also arise from top-down decoherence and subjective selection of the preferred basis. Box 6.3

develops theory with respect to the reallocation of consciousness across realities from time to time; this action, that might be seen as an occasional correction of previous decisions, is enhancing free will, but, perhaps, as will be discussed, at the price of having to change memories, to be called 'quantum brainwash.'

Chapter 7 then concentrates on the philosophical discussion around responsibility, partially building up upon Chap. 6. However, whereas the possibility of executing free will can unambiguously be shown in the clustered-minds multiverse, responsibility will turn out to be a subtler issue. Short-term, i.e., in each singular decision, *full* responsibility of a certain version of an individual for his actions is difficult to justify since a *complete* removal of consciousness from non-preferred realities is not possible. The chapter therefore brings in Buddhist' and Confucianist' accounts of responsibility that help establishing the idea of a *long-term responsibility* in the clustered-minds multiverse. Another potential limit for responsibility to be looked at in Chap. 7 are the many choices that are (partially) driven by unconscious motives. Chapter 7 contains one box. Box 7.1 discusses a pragmatic, economic approach to punishment (potentially replacing 'moral' responsibility with 'deterrence').

A step towards theory development in the decision sciences will be taken in Chaps. 8–10 (Part III of the book). All these chapters discuss potential problems that the multiverse perspective might pose for that discipline. The normative model of choice (i.e., expected utility theory) builds up upon the notion that a rational individual assigns utility levels to possible outcomes of choices and weighs them with the respective probabilities of their occurrence; the result is the expected utility for each alternative. Then, in a singular universe, a rational decision maker is supposed to choose the alternative with the highest expected utility. How could one translate this concept into the multiverse, is this straightforward, or must anything be changed in the decision-theoretic framework? Chapter 8 investigates this question for the general framework of normative decision theory (including its objective function) as well as for the aspect of probability. Whereas surprisingly, hardly anything has to be changed with respect to probability, as explained in Box 8.1 (it is only the view on probability that has to be changed, compared to the standard approach to decision making), the framework is seriously affected by having to replace the singular outcome of a choice in a singular reality with the vectorial outcome in a multiple-reality setup. The concept of *vectorial choice*, underlying the development in all remaining chapters of the book, will thus be introduced and discussed in this chapter. The general framework of normative decision theory will also be confronted with results from behavioral decision theory as well as the alternative framework of the *effectuation* principle (Sarasvathy 2001) that turns out to have some advantages for usage and further development in the multiverse.

Chapter 9 concerns itself with another aspect of the objective function for selecting between alternative realities: What makes individuals prefer one over another reality, or more precisely, one vector of 'reality weights' over another? How should the utility that is generated from a set of 'movies,' a set of realities that an individual has chosen, be defined? Is the appropriate concept of utility even clear for a singular reality? Is there anything to be learned from the concept of Bentham utility (i.e., experienced utility) or from Daniel Kahneman's more recent research on this matter (e.g., Kahneman et al. 1997)? And is it correct to assume that individuals are choosing in favor of the most positive developments? Aren't they also interested in making all kinds of odd experiences? Those considerations in Chap. 9 also play a role in Chap. 11 where—among other aspects—a 'neurotic' behavioral motive: repetition compulsion, is dealt with. Chapter 9 furthermore analyzes the appropriate utility concept for usage within the effectuation framework and the potential of making use of a multidimensional utility framework in the multiverse. It becomes evident that it is all but clear what type of utility is appropriate for usage within choices in the clustered-minds multiverse. Chapter 9 is rather able to give an account of the problems to be solved in future research.

Chapter 10 then looks at strategic and economic decisions. Specifically, a vast part of the chapter will be concerned with a certain form of a simultaneous market entry game that turns out to be extraordinarily interesting from the perspective of the multiverse because it is characterized by *multiple Nash equilibria* and hence by substantial uncertainty on the side of the players as to what strategy to choose (Rapoport 1995).[14] The literature has labelled some of the findings in those games as 'magic' (Kahneman 1988), because the respective authors could not make much sense of the remarkable coordination success in those games applying singular-reality theories. The chapter will show that some of the 'magic' findings, e.g., some of the experimental findings by Rapoport (1995), become reasonable when applying a clustered-minds multiverse perspective, and that those findings are informative in turn for the development of the theory of the clustered-minds multiverse—by helping to better understand the clustering by individuals. Experiments on other market entry situations (i.e., Camerer and Lovallo 1999) as well as games against computers will also be analyzed through the lens of the multiverse.

Chapters 11 and 12 discuss potentially far-reaching consequences of the clustered-minds multiverse for psychological phenomena as well as experimentation in the social sciences (Part IV of the book). During some undergraduate classes in psychology, many years ago, I have been confronted with basic concepts from psychoanalysis such as repetition compulsion—as most others studying psychology. I have since then asked myself how a neurotic person would be able to 'manipulate' her environment so successfully as to always get the 'right' people involved in their lives doing the 'right' things to them so that certain experiences *can* be repeated; please note that this phrasing is not meant to be cynical—the author is aware of how painful certain repetitions are for the individuals experiencing them. So, without wanting to take away anything from the sadness of those experiences, an individual having experienced a distant mother will unconsciously 'manage' to experience relationships with unloving individuals quite frequently in

[14]The economists and game theorists may apologize the 'sloppy' usage of terms in this introductory chapter, especially in light of the fact that mixed strategies also exist and are unique. The terminology will be used in a more precise manner in Chap. 10.

his life. How is that possible? I have finally found the answer, but in the multiverse. Consciousness is putting more emphasis on specific realities out of a plethora of possible realities, but that by no means implies that it always 'picks' those that are 'best.' Indeed, it is quite plausible that it sometimes picks those realities that are meeting conscious or unconscious 'expectations.' Individuals suffering from certain types of neurosis (but also non-neurotic individuals to some extent) expect life to treat them in a certain way. They will then allocate most of their consciousness to realities where specific versions of other individuals are willing to meet their expectations. (The reader will become more used to this type of reasoning throughout the book; see, e.g., Chaps. 4 and 6.) The chapter continues with the exciting phenomenon of self-fulfilling prophecies in several spheres of life and will discuss explanations for those phenomena that have been proposed in the literature, e.g., multiple equilibria in game theory and economics. The clustered-minds multiverse will turn out to offer a general framework containing those phenomena as well as some of the explanations that have been suggested for them.

If consciousness allocates more or less weight to certain measurement outcomes in a quantum experiment, why shouldn't it be able to allocate, say, more weight to certain measurement outcomes in a more general way? The entire book is going to deal with such phenomena: The whole point of the multiverse is that there are different versions of reality, and I will have argued in several chapters preceding the twelfth, that different individuals might opt for preferring, via a higher degree of consciousness allocated to them, different realities. The Rosenthal effect (Rosenthal 1976) is well known. It implies that a researcher, convinced of his hypothesis, might unconsciously influence the outcomes of his experiments in a way consistent with this hypothesis. Rosenthal has considered this being 'physical' (i.e., material) influences or he was purposely vague about this, talking about 'subtle cues.' E.g., the researcher influences the individuals carrying out the experiment in the laboratory, and they *somehow* influence the behavior of the participants. The *generalized Rosenthal effect* introduced in Chap. 12 is more fundamental. It claims that consciousness puts more emphasis on realities containing specific measurement outcomes, e.g., those where the outcome is consistent with the expectations.[15] A thought experiment (that might, perhaps with some modifications, be made an actual experiment) will be proposed that tests for such effects. Potential consequences for scientific research are outlined.

The last part of the book (VI: Conclusions and General Perspectives) contains only one final chapter (i.e., Chap. 13) that is concerned with consequences of the clustered-minds multiverse for our weltanschauung and for future research in physics and in the social sciences in general. This chapter does not aim at the impossible: summarizing from all the previous chapters the plethora of research opportunities that have been mentioned or the several consequences for our weltanschauung that arose. Instead, it focuses on some especially important and

[15]Since consciousness might cluster in 'meaningful' ways (see, e.g., Chap. 10), this scientific result will then 'inform' a certain minds cluster, it becomes scientific knowledge within that cluster.

challenging new aspects in this regard (or more detailed accounts of only briefly mentioned, but rather important phenomena). It wants to be stimulating instead of suggesting any completeness. Part of those challenging thoughts are the limits put on the potential revelation of 'truth' in scientific research in physics and in the social sciences by the clustered-minds multiverse, the introduction of a new type of psychophysical experiments, and, quite generally, the move from a quasi-Newtonian to a multiverse worldview.

A quick note on gender neutrality: I found it hard to write this text in a gender-neutral way, partially because I am not a native speaker of the English language. Also, gender-neutral language typically leads to a higher complexity of the sentences. Since the level of abstraction is already high at times, at least the examples, using real decision makers, where aiming at using an easy-to-comprehend language. Thus, when the text is talking about the decision maker as "he" or "him" etc., this is not meant to be offensive.

PART I
SETTING THE STAGE FOR THE CLUSTERED-MINDS MULTIVERSE

Chapter 2
Why the Multiverse Is the Most Parsimonious Way of Interpreting Quantum Mechanics

The 'Issue' with Quantum Mechanics: Defining the Measurement Problem

> I think I can safely say that nobody understands quantum mechanics.
>
> Richard Feynman (1967)

Has the situation substantially changed since Feynman stated the above in the sixties of the last century? I—at least partially—doubt it. On the one hand, quantum mechanics is a physical theory that makes very accurate predictions. On the other hand, people still disagree on how to interpret it in terms of the reality it describes. While some physicists then traditionally deny the *necessity* to understand it as long as it works, Feynman *not* being one of them, this position is quite unsatisfactory from a more conceptual viewpoint. The main problem of understanding quantum mechanics circles around the so-called *measurement problem.*[1]

In fact, the situation is quite strange. As long as *no* measurements are conducted, a quantum system is well described by the linear Schrödinger equation

[1]Great introductions, albeit from different perspectives, to the measurement problem are to be found in, e.g., Squires (1994), Barrett (1999), Auletta (2001), Mensky (2010), Wallace (2012c). Whereas Barrett (1999) applies a realist perspective and is neither articulated as being pro or as being con the multiverse perspective, Mensky is a proponent of a *subjective version* of the multiverse perspective. Squires is a bit more balanced, but also with a strong twist towards the multiverse interpretation of quantum mechanics; and he is also proposing to take into account subjective elements. Wallace 2012c is a proponent of the multiverse, albeit a realist version of it (the so-called Oxford interpretation; see below). Auletta introduces into the formalism of the different interpretations.

© Springer Nature Switzerland AG 2018

C. D. Schade, *Free Will and Consciousness in the Multiverse*,
https://doi.org/10.1007/978-3-030-03583-9_2

(Schrödinger 1926). This is the basic equation of quantum mechanics: a *wave function*.[2] But if we *measure* anything[3] on some quantum system, either with the help of a physical apparatus or just 'using' our human perception, this wave function *seems* not to be the appropriate description for what one gets.

Puzzling at first sight, but explicable within the framework of the wave function via *decoherence* (Zeh 1970; Zurek 1991) is the appearance of seemingly 'classical' measurement outcomes in the macro sphere. The most important departure, however, is that the wave function indicates *superpositions* of *different possible states* of the quantum system. And those superpositions are kept if the entanglement between measuring instrument and quantum system (but no further operations such as calculating a reduced density matrix) is considered (called *decoherence I* below).[4] Simplifying, a superposition indicates a situation where *alternative states* that would exclude each other from a 'classical' perspective *coexist* (see the illustrative Schrödinger's cat example below). But at the same time, "(…) we know that it is only one of the alternatives, not all of them, that is observed in any real experiment. This is the main conceptual difficulty of quantum mechanics leading to what is called 'measurement problem'" (Mensky 2000b, 24).[5]

A simple case for the fact that the problem of superpositions on the one hand and potential, definite measurement outcomes on the other hand might propagate to the *macro world* was made in a famous thought experiment, known as *Schrödinger's cat*:

---

[2]Note that part of the empirical support for the wave function is indirect, based on a stochastic calculus based on this function (Born 1926). And this might already be seen as the beginning of the trouble since the wave function is deterministic. But there is also direct support for the fact that 'particles' are well described as being waves. An important example is the double-slit experiment where such 'particles,' e.g., photons, in fact *narrow wave packets* (see, e.g., Wallace 2012, 65; Zeh 2016a), are fired towards a plate pierced by two parallel slits. Behind the plate there is a screen that measures what ends up there. With both slits open, an interference pattern occurs (see, for a great description of this experimental paradigm, Barrett 1999, 2–8). The original double-slit experiment demonstrating the wave-like nature of light has been carried out first by Young in 1803, already; the first experiment of this type using electrons has been designed and carried out by Jönsson (1961). The double-slit experiment typically continues by measuring the path the photon was taking (which-way experiment), and in the traditional setup of this (gedanken-) experiment, the interference pattern then disappears. This has traditionally been interpreted as waves becoming particles; but if 'particles' are, anyway, narrow wave packets, the explanation must be different.

[3]An exception are, to some extent, clever experiments employing certain measurements carried out at entangled particles to measure the way the 'particle' of interest is taking (which-way experiment) whilst keeping the interference pattern in the double-slit experiment (see below).

[4]If one either calculates such a reduced density matrix (see the below discussion) or assumes a collapse of the wave function, one leaves the regime of the wave equation.

[5]The fact that in practical applications of quantum mechanics one would normally first calculate an ensemble of possible measurement outcomes (or quasi-classical realities) that occur with classical probabilities given by the Born (1926) rule or via the calculation of a reduced density matrix (see below) should not distract from the fact that predicted realities are multiple but perceived only singular.

> One can even set up quite ridiculous cases. A cat is penned up in a steel box, along with the following diabolic device (...): in a Geiger counter there is a tiny bit of radioactive substance, *so* small, that perhaps in the course of one hour one of the atoms decays, but also, with equal probability, perhaps none; if it happens, the counter tube discharges and through a relay releases a hammer which shatters a small flask of hydrocyanic acid. If one has left the entire system to itself for an hour, one would say that the cat still lives *if* meanwhile no atom has decayed. The first atomic decay would have poisoned it. The (...) [wave function] would express this by having in it the living and the dead cat (pardon the expression) mixed or smeared out in equal parts. It is typical of these cases that an indeterminacy originally restricted to the atomic domain becomes transformed into macroscopic indeterminacy, which can then be resolved by direct observation. (Schrödinger 1983 [1935], 156)

As has already been mentioned (effects of decoherence), the relevance of the quantum 'indeterminacy' for the macro world of our life is not restricted to *constructed* cases such as Schrödinger's cat. It is a general phenomenon. It will also become clear that, in narrow terms, it might *not* be an actual 'indeterminacy' that we are confronted with. Furthermore, the entire quantum mechanical calculus appears to be relevant to the macro sphere, not only the seeming indeterminism. However, "(...) [until] recently, one did (...) generally believe that some conceptual or dynamical borderline between micro- and macrophysics must exist—even though it could never be located in an experiment" (Zeh 2016a, p. 202).[6] A typical research strategy in this regard is to show quantum effects with larger and larger units such as a mesoscopic paddle of the size of 30 micrometers (O'Connell et al. 2010). There is no good reason as to why decoherence should stop at any size of objects.

Some real experiments, carried out on the properties of spin-1/2 'particles' in the micro world, provide more evidence for the *weirdness or "Alice-in-Wonderland logic"* (Lockwood 1996, 160) *of quantum measurement*. (Talking about 'particles' is always an idealization since 'particles' are in fact narrow wave packets; see, e.g., Wallace 2012, 65; Zeh 2016a.[7]) The basic structure of the following description is similar to that underlying the presentation in Barrett (1999, 9–11). As in the respective part of the monograph of this author, the following analysis will focus on the case of *electrons* that are one case of spin-1/2 'particles.' A nice figural explanation as to what a spin is has been provided in Lockwood (1996):

> Spin is an intrinsic angular momentum which, like ordinary angular momentum, can be oriented in any spatial direction. Think of an electron as being pierced by an arrow aligned with the axis of spin. Then the electron is said to be *spin-up* if, as viewed by an imaginary observer looking along the arrow from its tail, the spin is clockwise, and *spin-down* if it is anticlockwise. (160)

---

[6]The reason as to why there is no such borderline is complex, multiple decoherence, i.e., quantum correlations (entanglement) that span between the micro and the macro sphere (Zeh 2016a; Mensky 2010; Mensky 2000b).

[7]This view has already been expressed within a few above footnotes. I am aware of the fact that this view is in contradiction to Heisenberg's view; it is, however, fully consistent with the view by other physicists such as Mensky, Zeh etc. I will come back to this issue, below.

Let me look at the *x*-spin and *z*-spin property of an electron—they are related in interesting ways (a more detailed exposition of this textbook example can be found in Barrett 1999, 9). Both types of spin can be *up-spin* and *down-spin*. To start with, electrons (as any other spin-1/2 'particle') are in a *superposition* regarding the up- and down-spin properties of those spins. A measurement of either spin generates the same (definite) result when repeated, as long as nothing happens in-between the two measurements. However, if

> (...) one finds that an electron is *x*-spin down, and then measures its *z*-spin, one gets each of *z*-spin up and *z*-spin down about half of the time. If one in fact gets *z*-spin down and *remeasures* its *x*-spin, one gets each of *x*-spin up and *x*-spin down about half of the time. (...) the result of a subsequent *x*-spin appears to be completely random regardless of how careful one is making the intervening *z*-spin measurement. Similarly, *x*-spin measurements appear to randomize completely the *z*-spin of an electron. (Barrett 1999, 9)

In other words: Whenever we know one of the two spins, we have no knowledge about the second spin. But if we then measure the second spin, the knowledge of the first spin is destroyed. In line with Barrett (1999, 9), I would like to argue that this is inconsistent with any 'classical' idea of measurement in our macro world.

To understand better the difference of measurements carried out in the classical world and in the quantum world, let me look at the following straightforward example: If I have first measured the air temperature in my office and then measure, say, the humidity of the air, would my knowledge of the temperature be destroyed? Certainly not.[8] So the odd outcomes of the above spin measurements allow us an insight into the strange world of quantum measurement where other rules appear to apply than in our 'regular environment.'[9] I would like to, furthermore, mention that the oddities of quantum measurement in this first experiment are appearing in a slightly different structure than that usually employed to experimentally demonstrate the measurement problem.

A structure more typical for the demonstration of the measurement problem can be realized in the second part of Barrett's analysis of the textbook example (again, see Barrett 1999, 9–11, for a more detailed exposition) which is concerned with another type of experiment. Of course, the effects that lead to the following results partially resemble those that are at work in the first experiment. In the second experiment, electrons are sent into an *x*-spin sorting device where, depending on the *x*-spin of the electron, it will take *one of two different paths* that both end up, however, in the same place.

---

[8] I am abstracting here from small changes in the room's temperature caused by my measuring activities.

[9] I would also like to underline Barrett's thought that this 'conspiracy' somehow resembles the *Heisenberg uncertainty relation* for position and momentum of particles where only one of the two can be known at a time (Barrett 1999, 9, especially footnote 3).

> Suppose we send z-spin up electrons into the x-spin sorter: what should the statistics be for a z-spin measurement at (...) [the final place]? One might expect about half of the electrons to take (...) [one of the paths] and about half to take (...) [the other path]. (Barrett 1999, 10)

What we now observe at the final measurement point, however, depends on whether or not we *look* at the two paths. If we do look, we will find half of the electrons travelling along one of the paths (those with x-spin up), and the other half (those with x-spin down) taking the other route. We would also find, consistent with the first experiment, half of the electrons at the final measurement point (where they all 'meet') to exhibit z-spin up and the other half z-spin down. If we do *not look* (i.e., do not measure anything at either of the paths), the electrons end up with the same spin they started off with, *all* exhibiting z-spin up!

So, if we do not look at the path, the behavior of the 'particles' is consistent with the 'particles' being (and remaining) in a so-called superposition between x-spin up and x-spin down states and it remains fully unclear which path they were taking at any point. A verbal description of such a superposition situation might be provided in the following way (see, identically, but for the case of the double-slit experiment, Barrett 1999, 5): A 'particle' with a superposition between x-spin up and x-spin down does not determinately pass through the x-spin down path, does not determinately pass through the x-spin up path, does not determinately pass through the x-spin down path and determinately pass through the x-spin up path, and does not determinately not pass through the x-spin down path and determinately not pass through the x-spin up path. Only this way, the fact that all electrons finish in z-spin up can be explained. However, when the electrons are *looked* at and their x-spins actually measured along each path, not only this superposition appears to somehow vanish, but a totally different outcome of the experiment occurs, namely, half of the electrons at the final point are z-spin up and half are z-spin down. *Observing something seems to change the observed.*

People have debated different solutions to the measurement problem from the early times of quantum mechanics on. Each of the solutions draws a different picture of reality, or better: implies a different *weltanschauung*.[10] Wigner (1983) [1961] pointed out that measurement results achieved on some quantum system are generally uninterpretable without taking into account the *consciousness* of the observer. Conscious observation turns out to be the end of a logical chain of reasoning defining a measuring device, then defining the brain as evaluating the result shown on the measurement device, etc. The exact *role* that consciousness plays is, in turn, directly connected with the interpretation of quantum mechanics chosen.

Not everyone agrees with the perspective that consciousness is at the core of the solution of the measurement problem. Whereas this perspective plays an important

[10]Hence, the measurement problem is also an 'ideological' issue as has already been suggested in the introduction. The fact that different interpretations of quantum mechanics are likely leading to an entirely different *weltanschauung* (= world view) has proven not to be helpful, I think, in interpreting quantum mechanics in the last hundred years.

role in the theorizing by Squires (e.g., 1988) and Mensky (e.g., 2005, 2010) and will also be pursued in this book, Barrett (1999), e.g., is strongly opposed to this perspective:

> Wigner tried to solve the measurement problem by saying precisely what it is that distinguishes observers from all other physical systems: observers are *conscious*. (…) [But] is it really necessary to introduce something extra-physical (in this case minds) in order to solve the measurement problem?! (Barrett 1999, 55)

Unlike Barrett himself, this book will answer his question with a clear "Yes," as explicated below as well as in Chaps. 3, 4 and 5. And the answer offered in this book will be more radical than the quite 'passive' treatment of consciousness in the Oxford school literature on the multiverse interpretation (Wallace 2012; Saunders et al. 2012) or in the considerations by Zeh (2013).

The two most well-accepted interpretations of quantum mechanics are the *Copenhagen interpretation* (nowadays an 'umbrella term' for various single-world interpretations such as Bohr's and Heisenberg's, and, most importantly, for von Neumann's reduction postulate: von Neumann 1996 [1932])[11]) as well as the many-worlds interpretation, initially based on Hugh Everett (1957) and its further interpretation by Bruce DeWitt (1970, 1971) and since then 'interpreted'[12] in several ways—in this monograph called *multiverse interpretation.*[13] The Copenhagen interpretation of quantum mechanics is the one most representative of something that might be called a 'quasi-Newtonian' worldview; it is that interpretation of quantum mechanics that challenges the validity of our everyday experience in the least radical way. It leads to a singular reality with 'quasi-classical' properties. The opposite holds for the multiverse interpretation. The existence of a plethora of realities stands in sharp contrast to our everyday experience. The Copenhagen interpretation is underlying several practical applications of quantum mechanics. Whereas the multiverse interpretation is underlying many approaches in the area of quantum computing. Within the academic community *in physics*, the most vivid proponents of the multiverse are (or have been, respectively) David Deutsch, Murray Gell-Mann, Richard Feynman, Stephen Hawking, Michael Mensky, Don Page, and H. Dieter Zeh (please note the considerable overlap with the authors of the literature referenced in the section

[11]For simplicity, whenever the paper mentions the Copenhagen interpretation, this (most prominent) version of it is meant. So, I am using the term 'Copenhagen interpretation' as a synonym for the so-called 'standard interpretation' of quantum mechanics, although this might be a historically wrong classification. It appears to me, though, as if this were a convention that has evolved in large parts of the literature.

[12]The idea that the multiverse interpretation needs interpretation goes back to Albert and Loewer (1988).

[13]The use of the singular has to be justified, here, given the various interpretations of the multiverse interpretation (see Chap. 4) existing. Although differences between those interpretations will play a major role in this book, they all share the common characteristic of considering the parallel existence of different realities. It is hence justified to still see them as part (or varieties) of one interpretation.

"This Book's Contribution to the Literature" in the first chapter). In philosophy, however, the most well-known contemporary school dealing with the multiverse interpretation is the so-called 'Oxford school' including scholars such as David Wallace and Simon Saunders (again, see also the introduction chapter). Slightly older are the contributions by Michael Lockwood and Euan Squires. For the approach developed in this monograph, the seminal works by David Deutsch, Michael Lockwood, Michael Mensky, Euan Squires, David Wallace and Dieter Zeh turned out to be the most important.

Other interpretations and *modifications* of quantum mechanics have been proposed.[14] Of all those, the *objective reduction formalism* (Penrose 1994; Hameroff and Penrose 1995), e.g., will briefly be touched in this book. The same holds for the more well-known hidden variable theories such as the de Broglie-Bohm *pilot wave theory* (e.g., Bohm 1952). This theory, however, will not be dealt with in more detail for numerous reasons.[15]

Not an interpretation of quantum mechanics is the theory of *decoherence* (e.g., Zeh 1970, 2012; Zurek 1991, 2002; Mensky 2000a, b, 2001; Joos et al. 2003). Since it plays a large role in the literature on the measurement problem (as well as in the literature trying to avoid it), because it is important to understand the relevance of quantum phenomena for the macro world and because it is at the core of understanding the functionality of the multiverse, it will be dealt with in some detail below.

Copenhagen Versus the Multiverse

What are the most important differences between the two mainstream interpretations of quantum mechanics, is there experimental evidence favoring one over the other?[16] The Copenhagen interpretation proposes a 'collapse of the wave packet' when a system undergoes measurement (so-called reduction), i.e., a disappearance of all elements of the superposition—or of all potential, quasi-classical outcomes of the measurement, if transformed into an ensemble—except *one*. The multiverse interpretation states that nothing like this happens, that no reduction will ever take place. Whereas the Copenhagen interpretation *needs* the reduction postulate that is in contradiction to the universality of the Schrödinger equation,[17] providing an account of the appearance of a *singular* reality is relatively easy within this

[14]For an overview and detailed analysis see Auletta (2001).

[15]Among various issues, one issue with that theory is the following: "(…) there is one problem with the Bohm theory which we must mention, (…) that it is not compatible with special relativity" (Squires 1994, 83). Another is brought up by Polkinhorne (1984) who compares this interpretation with the epicycle explanation for planetary orbits (implying that it is perhaps tautological); this, however, is called a slightly unfair judgment by Squires (1994, 120).

[16]Detailed introductions into the differences between those two explanations are to be found in Barrett (1999, 18–91) as well as Mensky (2005, 2007a).

[17]In Mensky's words, the implied collapse of the wave function is 'alien' to quantum mechanics (Mensky 2005, 2007a, 2010, chapters 1 and 2).

framework, if the notion of a 'probabilistic selection' of one out of several alternative outcomes (and the idea of a collapse) is accepted to start with[18] (and the preferred basis problem disregarded; see Chap. 4). In fact, that might have been one of the reasons of coming up with the Copenhagen interpretation in the first place when quantum mechanics was new and possibly puzzling its inventors. On the other hand, coming up with an account of the appearance of a singular reality is somewhat *more complex* if the multiverse interpretation is adopted, since it leads to the consequence that somehow, different—and separate—observations are made in parallel within an observer's consciousness. Indeed, our reality is not unique, anymore, and "(…) [the] question then to be answered is why an observer sees only one alternative in practice" (Mensky 2000b, 24). However, the theory is fully convincing from a formal standpoint. Unlike the Copenhagen interpretation, most versions of the multiverse interpretation do not contain any problematic assumptions or postulates—as will be demonstrated in the book—since the wave function is assumed to be universal.[19] (Although the wave function is assumed to be universal, the Born (1926) rule is still required—in my understanding an auxiliary equation; see Chap. 4—to generate relative frequencies of measurement outcomes within one reality—along one time line, if you will—and to thus provide a pragmatic coupling between measurement outcomes and the wave equation.)

As argued in many parts of this book, it appears to be hard to generate clear-cut experimental evidence within physics that can be interpreted in favor of either the Copenhagen interpretation or the 'many-worlds' view.[20,21] In his monograph on "Quantum Measurements and Decoherence: Models and Phenomenology," one of his earlier works on this matter, Mensky (2000b), even states:

[18]Most certainly, Einstein did not accept this type of reasoning. In a letter that Einstein wrote on December 4th, 1926 (to either Max Born or to Nils Bohr; this not quite clear), he said (my own translation from German): "Quantum mechanics is very awe-inspiring. But an inner voice tells me that this is not the final thing (nicht der wahre Jakob). The theory delivers a lot, but it does not bring us any closer to god's secret (dem Geheimnis des Alten). In any case, I am convinced that He does not play dice."

[19]More details on problematic assumptions in some versions of the multiverse interpretation will be provided in Chap. 4.

[20]The situation is unclear enough that David Deutsch and Michael Mensky, two vivid proponents of the many-worlds view on the physics side, disagree on the evidence presented within physics. Deutsch believes that the experimental evidence generated within physics is already in favor of the many-worlds view (Deutsch 1997, Chaps. 2 and 3). He even *identifies* quantum mechanics with, how he calls it, the *Everett theory* (Deutsch, 2012a). However, Mensky (2005, 2007a, b, 2010) argues that the evidence generated within physics cannot unambiguously be interpreted in favor of the multiverse view.

[21]Different authors use different terms for the multiverse interpretation(s), and it would be difficult to always replace the respective authors' terminology by mine, especially within references. Therefore, some terms will be used synonymously, such as 'many-worlds' interpretation and multiverse interpretation. When the topic requires to be specific, i.e., to discriminate between 'many-worlds' and 'many-minds' interpretation, I will do so.

> (…) [The] strange picture of parallel worlds is logically quite possible. Whether it solves the [measurement] problem is for the reader to decide. There is an opinion that it solves nothing. One has always to ask whether a theory can, in principle, be falsified, that is to say, whether it is testable. Referring to Everett's many worlds interpretation it seems to be impossible to verify the existence of many worlds (corresponding to the many alternatives of a quantum measurement) by methods used in science. In this sense the question does not lie within the scope of physics. (25)

Although no direct test of the multiverse interpretation, *Bohr's complementarity principle*, however, closely related to the Copenhagen interpretation and implying that physical entities may either behave as a 'particle' or a wave but never both ways at the same time, got more and more undermined by novel experimental findings at the double slit (for a short description of the double slit experiment in the context of the measurement problem, see Barrett 1999, 2–8; see also the footnotes accompanying the introductory section of the current chapter). By using very clever experimental designs, some research groups (see, e.g., Mittelstaedt et al. 1987; Scully et al. 1991; Menzel et al. 2012) have demonstrated that it is possible to partially or fully keep the interference pattern (wave-like behavior) whilst nevertheless measuring the path the respective 'particle' was taking. Whereas those findings are unfavorable for the Copenhagen interpretation, they are leaving the many-worlds interpretation untouched.[22] Indeed, in his 1997 popular science book "The Fabric of Reality," David Deutsch seems to indirectly base his argument pro the existence of a multiverse already on those novel findings. When discussing the interesting change of the interference pattern of a singular photon sent through four versus two slits even though the way of the photon through one of the slits can clearly be identified (Deutsch 1997, Chap. 2), he leaves the possibility unmentioned which has been postulated as the outcome of a *gedanken* experiment early in the history of quantum mechanics (and often been demonstrated in actual experiments, later): that measuring the path of a photon in a double-slit experiment ('which-way' information) would *destroy* the interference pattern; this gedanken experiment was developed in close connection with the complementarity principle stating that photons might either behave as a wave *or* as a 'particle.'

Sure enough, those novel findings at the double slit have not necessarily been interpreted in favor of the many-worlds view by other physicists. Just one, perhaps quite unspectacular example are poster and paper by Maria Cruz Boscá (2007, 2009)[23] who discusses, in light of those novel findings, the necessity to change the formalism of quantum mechanics, to formulate further assumptions, to modify the complementarity principle etc. Nothing more 'radical' is mentioned.

Perhaps, all those findings on the particle-wave duality as well as David Deutsch's description and interpretation of the findings at the double/quadruple slit must be seen somewhat critically if utilized as empirical evidence in favor of a certain solution to quantum mechanic's interpretation problem, at least without any

[22]For another (still hypothetical) way of potentially discriminating between different interpretations of quantum mechanics, see Deutsch (1985).

[23]The poster is referenced with the permission of the author.

further qualification. If one takes the universality of the wave function as given and considers quantum correlations (entanglement) as the only permitted type of manipulation, *there are simply no particles* out there (as postulated in the Heisenberg picture), but only *narrow wave packets* appearing as if they were particles (e.g., Zeh 2016a). Zeh reported to have been actually quite confused when he saw at a conference, many years ago (Varenna conference of 1970), "DeWitt translate Everett into the Heisenberg picture. For (…) [him], Everett's main point was a unitarily evolving wave function of the universe" (Zeh 2016). This view is shared by another proponent of the multiverse, Michael Mensky (e.g., Mensky, 2005, 2007a, 2010). And I personally agree with Zeh and Mensky. (An important deviation from the wave equation, however, is brought up by Mensky in connection with his free-will analysis to be discussed in Chap. 4.)

In this book, however, I do not attempt to reconcile the Heisenberg and the Schrödinger pictures of the (quantum) world. I will rather rely on the Schrödinger picture whenever possible, thus staying close to Everett's (1957) original view (as well as Mensky's and Zeh's). But I will also follow a pragmatic route in so far, as some results generated within the Heisenberg framework are potentially important. Hence, I will bring in thoughts and theories developed within the Heisenberg picture when appropriate; and whenever I do so, I will point out that those thoughts have been generated within a different (potentially incompatible) framework and critically discuss them, if possible. The latter is far from trivial for the wave-particle duality and the double/quadruple slit experiment and will not be possible within the scope of this book.

As already stated above, in this monograph I will follow the lead of Wigner (1983) [1961], Squires (e.g., 1988), Mensky (2005, 2007a) and Stapp (2009) in the conviction that consciousness *always* plays a central role in the *solution* of the measurement problem, no matter which interpretation of quantum mechanics will be adopted.[24,25] In fact, the entire interpretation problem is reaching another level, philosophically, when the always central, but clearly different role that consciousness plays in the Copenhagen and in the multiverse interpretation is analyzed. As already dealt with, quantum systems, including the measurement device etc., are in a superposition state with alternative states 'coexisting;' and the main question then is why consciousness *perceives* only *one* reality (see also the above thoughts by Mensky). Since, from the perspective of all authors referenced at the beginning of this paragraph, a unique result or outcome of the measurement is determined *only* when consciousness of the observer comes into play, paradoxes can be constructed such as Schrödinger's cat (see the beginning of this chapter) or *Wigner's friend.*

[24]Consciousness is here identified with *qualia*, the 'hard-problem' aspect of consciousness as sort of a *pure subjectivity* (see Chalmers 1995, 1996). For more details on the definition and role of consciousness see Chaps. 3, 4 and 5.

[25]There might be philosophers and physicists pursuing a realist interpretation of the multiverse that disagree with me on this. And I would like to argue that they are missing out on something important in their theorizing as will become clear throughout this book. Therefore, the way to proceed in this book will be to substantiate the subjective, consciousness-based account as much as possible.

In Wigner's friend (Wigner 1983 [1961]), a friend of the principal investigator conducts a measurement at some quantum system for him, e.g., measures the outcome of Schrödinger's cat experiment, whilst the principal investigator is absent from the laboratory. The question is when the outcome of the measurement is actually determined, only after the return of the principal investigator, or at a previous moment, i.e., when the friend has looked at the device but the principal investigator is not informed about the outcome, yet?

According to the Copenhagen interpretation, something *discontinuous* takes place whenever conscious observers conduct measurements. Since the wave function is assumed to collapse, consciousness might be seen as responsible for this collapse (Stapp 2009). Clearly, this is the most straightforward way of making sense of the Copenhagen interpretation. Where else would the collapse of the wave function be 'executed:' in the measurement device, in the retina or in the brain? Nothing may make any of these stages 'special' enough. Therefore, consciousness must always be the end of a stepwise regress.[26] Note that this is a modern perspective. The question where the 'cut' between the regime of the wave function and the regime of the 'quasi-classical' world (after the assumed collapse) is to be made has been debated for decades and is what is referred to as the location of *Heisenberg's cut*: Born and Pauli located the cut directly after the to be measured quantum system, Bohr had it located somewhere in the macro system, and only Wigner then moved it into the domain of consciousness (see Zeh 2012, 77–84).

Everett (1957) finally moved Heisenberg's cut even farther away to a point where never any collapse occurs.[27] In the multiverse interpretation, the exact action of consciousness is interpreted differently depending on the *subtype* of this interpretation adopted (the analysis of those differences and their consequences is at the core of Chap. 4). However, the common element of *all* approaches under the heading of the multiverse interpretation indeed is that consciousness is *not* collapsing anything since the wave function is universal. Then, according to the perspective pursued in this monograph (also to be explicated in detail in Chap. 4), consciousness might prefer certain realities over others and might put more *emphasis* on the preferred out of the vast number of possibilities out there. Despite all the substantial differences, some authors in the paradigm of the Copenhagen interpretation and some subtypes of the multiverse interpretation (including the one pursued in this book) share the view of consciousness being at the center of a process of *'creation' of subjective reality*.

[26]Somewhat related to this, based on the objective reduction formulation of quantum mechanics, there are approaches that link this assumed collapsing action of consciousness to observable processes in the brain (e.g., Hameroff and Penrose 1995; Hameroff 2012).

[27]For this perspective on theory development in quantum mechanics see Zeh (2012, 79).

Lessons from Decoherence

"The phenomenon of decoherence occurs whenever [a] (…) system interacts or is made to interact with its environment, and the state of the system has some impact on the environment"[28] (Mensky 2000b, 5). The basis of decoherence is *entanglement* (Mensky 2000a, 592; see also Zeh 1970; Zurek 1991, 2002). This means that the system and its (measuring) environment lose their independence in a fundamental way (become quantum correlated). It *theoretically implies* the well-known—and already mentioned—fact that according to quantum mechanics, measurement changes the measured. The mathematical decoherence apparatus involves two basic steps[29]; whereas the first, from here on: *decoherence I*, naturally follows from the Schrödinger equation, the second, from here on: *decoherence II*, is more pragmatic, and this turns out to be important.

It is often argued that decoherence (the full process involving steps I and II) *solved* the measurement problem because it explicitly models the entire process from quantum reality (i.e., the Schrödinger equation) to a 'quasi-classical' reality (macroscopic objects appearing not to be in superposition).[30] Clearly, this is not quite true. First of all, the end point of decoherence analysis, stage I and II, is always a *mixture* of alternative states (an ensemble with classical probabilities!). It is still unclear how a *singular* reality appears: Randomly, as the classical probabilities, identical to those generated within the Copenhagen interpretation, seem to suggest (despite the deterministic nature of the Schrödinger equation!), or via some action of consciousness? Moreover, the process of decoherence is misunderstood, resulting from the second, *pragmatic step* of the mathematical apparatus (decoherence II), and then *using* this very apparatus to address philosophical questions. This problem has rarely been spelled out in a satisfactory way, with the exception of Zeh (2012, 77–84). Since the issue is very important for the understanding of the measurement problem, I will go into some detail, here.

Decoherence I is simply calculating the entanglement between measured system and measuring device and nothing else; therefore this step is fully consistent with the Schrödinger equation. In the language of quantum mechanics, the system moves from a factorized state (where the two systems are independent) to an entangled (i.e., quantum correlated) state (see also Mensky 2000a, 592). The entangled state is still a *pure state*:

[28]Mensky dates back the idea of decoherence to Heisenberg. But it is really Zeh (1970) who presented the first formal treatment of decoherence. It was first called decoherence by others (see Zurek 1991).

[29]This is not meant to imply that there are only two calculations to be made.

[30]According to Mensky (2000b), there is an important aspect leading to 'macroscopically distinct' states with macroscopic measurement devices and to the orthogonality assumption of decoherence analysis being justified: "(…) [The] corresponding wave functions depend on very many variables and exhibit different functional dependence on the large number of these variables. The scalar product of such wave functions is practically equal to zero" (19).

> This essential and uncontroversial first stage of the measurement process can be accomplished by the means of the Schrödinger equation with an appropriate interaction. It might be tempting to halt the discussion of measurements with [the correlated state] equation. (...) Why ask for anything more? (Zurek 2002, 7)

However, this part of the mathematical analysis does not deliver the *states of nature* we are able to measure or perceive in the macro sphere, it does not generate the appearance of a quasi-classical reality (see also Zeh 2012, 82):

> The reason for dissatisfaction with [the correlated state equation] as a description of a completed measurement is simple and fundamental: In the *real world*, even if we do not know the outcome of a measurement, we do know the possible alternatives, and we can safely act as if only one of those alternatives has occurred. (Zurek 2002, 7; italics mine)

And here the above-mentioned pragmatism comes into play. The decoherence II calculus now *implements* our perception of a classical reality in measurement theory, and it makes a couple of assumptions, some more, some less problematic: "If we are only interested in the state of the system (…) (and not its environment (…)), we can describe this state by the so-called reduced density matrix" (Mensky 2000a, 592). Everyone takes this step in decoherence analysis, and it seems to be a straightforward, obvious step to take, and it is in fact satisfactory for all practical purposes. In fact, it is an alternative to the application of the Born (1926) rule, needed to generate relative frequencies of quantum measurement outcomes along one 'history' (see Chap. 4). Whereas, as already mentioned, I am going to call the Born rule an auxiliary equation, the decoherence II stage might analogously be called an *auxiliary stage*. And as the Born rule will turn out to be an important auxiliary equation, part of quantum calculus (see, again, Chap. 4), I am by no means saying that the decoherence II stage is unimportant, from a practical/empirical perspective.

However, it is *the* problematic step, according to Zeh (2012, 82), if one wants to interpret measurement via decoherence. It is leading to a deep theoretical misunderstanding of the theory of decoherence. What exactly *is* problematic is the fact that the density matrix *does not allow to discriminate* between entanglement (wave function) and a mixed state—regular probabilities of occurrence of the macroscopic measurement outcomes just mentioned (Zeh 2012, 83). But it has introduced this Copenhagen-type 'quasi-classicality' through the backdoor—or better *implicitly assumed* it via a pragmatic mathematical calculus. Hence, using the outcomes of the reduced density matrix calculus to state that the measurement problem is solved is a highly problematic argument. According to Zeh (2012, 83), it is perhaps acceptable to *pretend*, for pragmatic reasons, that a collapse into one of the measurement outcomes had taken place, but from a theoretical standpoint one should know better. *Decoherence does not solve the measurement problem.*

Another, even more central aspect with respect to the philosophical interpretation of decoherence is the fundamental difference between measurements carried out on closed systems (effectively the entire universe) versus measurements carried out on open systems (Mensky 2000a, b, 2001):

> If the measurement is described in terms of an open system, this system can be as broad as desired, but there still have to be some degrees of freedom outside in which information about the result of the measurement will be recorded in one way or another. This condition is characteristic of quantum measurement, and corresponds to the well-known arbitrariness in dividing the entire Universe into the measured system and the measuring instrument. (Mensky 2000a, 594)

But if analyzing open systems is arbitrary as Mensky points out, what would happen, including the effects on the phenomenon of decoherence, if we do move from some open system to a closed one (the entire universe)? This was indeed one of the core issues that Everett tried to address and that lead him to his multiverse interpretation. Everett's approach "considers the closed system that includes the measured subsystem, the measuring device, the observer—in short, all the Universe, the whole world. Accordingly, there is no decoherence" (Mensky 2000a, 595). This implies that, *objectively*, decoherence does not exist. It only appears in the subjective view of the world. And that does not only apply to the problematic, within a conceptual view (or approximate, if you will), calculation of mixed states.[31]

It appears as if something similar to the question of Heisenberg's cut that was discussed with respect to the Copenhagen interpretation earlier reappears when questions of decoherence are carefully analyzed within the framework of the multiverse. And it was actually Heisenberg, who "invented a historically unprecedented role for 'human observers.' He assumed that properties of microscopic objects are *created* in an irreversible act of observation—for him confirmation of the superiority of an idealistic world view instead of materialism, realism and reductionism" (Zeh 2013, 97; italics as in the original; see also Beller 1999, Chap. 4).[32]

---

[31]The last Mensky quote actually continues in the following way: "(…) and there is nothing to transform the superposition of alternative pure states into a mixture" (Mensky 2000a, 595), which is misleading. It might be read as if it required decoherence II with its artificial assumptions to move away from the world of the universal Schrödinger wave equation whereas Everett's (1957, 1973) relative state theory is consistent with the interpretation that it is really about the consequences of measuring inside an open system to get entanglement. It can be debated whether Everett really wanted to push a fully 'subjectivist' view. In fact, Everett was purposely using non-human observers (artificial intelligence) in his examples. But it is a negligible step to replace Everett's robots by conscious humans; and it is then also fully consistent with his theory to consider measurement inside the system as leading to *subjective* entanglement, only. But what about those processes leading to entanglement that seem to occur without any observer such as radioactive decay? Whereas I have to admit that this is really pushing the boundary of this argument (and not addressed in Everett's theorizing, or, to the best of my knowledge, elsewhere), I would still hold that it requires an inside-the-system observer to observe processes such as radioactive decay that are invisible for an outside observer of the universe. It should be noted here, again, that the Schrödinger equation is deterministic, indeed.

[32]I have promised to the reader to comment on results or standpoints that I am reporting in this monograph based on Heisenberg's matrix mechanics maintaining the idea of the existence of actual, quasi-classical 'particles.' Here, it is quite speculative to 'predict' whether Heisenberg would have argued in the same way if he had used the Schrödinger wave equation instead of his own matrix mechanics. Because many phenomena in regard to measurement are less surprising

Table 2.1 Quantum reality and classical reality in the multiverse when applying the theory of decoherence

Name of phenomenon	Type of 'reality'	Name of state	Mode
Superposition	Objective quantum world (closed-system perspective)	Factorized state	Objective all-is-possible-mode
Decoherence I: Entangled system in superposition	Subjective quantum world (open-system perspective)	Correlated state Relative state	Subjective as-well-as-mode
Decoherence II: Reduced density matrix	*Relative frequencies of quasi-classical states (within one 'history')*[a]	*Mixed state*	*Either-or-mode (along one time line, i.e., within one reality)*
'Selection'	Perceiving a singular, quasi-classical reality	Measurement result (Relative state)	Singular-outcome-mode (with one version of the individual)

[a]This phrasing foreshadows my pragmatic interpretation of the Born (1926) rule in chapter 4
Note The step depicted in italics might better be left out if conceptual problems are analyzed; more details on this step are provided, indirectly, in conjunction with the discussion of the Born (1926) rule in Chap. 4

Generally, multiple such (potential) Heisenberg cuts can be derived from the overview in Table 2.1 that is depicting the way from quantum reality to experienced quasi-classical reality from a multiverse perspective and taking into account the decoherence process. The first cut lies between objective quantum reality in a closed-system perspective (of an outside observer of the universe) and the subjective quantum reality (of an inside observer of the system) in an open-system perspective. Whereas Everett's (1973) *relative states* are argued by him to already occur with measuring and reporting robots (his example of observers), I would like to hold that it requires conscious observers (albeit split into different versions) to achieve subjective entanglement with certain states of the system (see also Chap. 5).

The second cut occurs between subjective entanglement and the reduced density matrix. This step *re-Copenhagenizes* the process and introduces some sort of 'collapse' through the backdoor. Despite its predictive accuracy (and necessity, as an alternative to the Born (1926) rule, if one wants to analyze relative frequencies of measurement outcomes within one 'history;' see Chap. 4), decoherence II might better be left out from the analysis if *conceptual* problems are analyzed.

The last step then is the 'selection' of the reality to be perceived by the observer. In the multiverse, this is not a selection in the sense of one reality appearing and all others 'disappearing,' however. It is also not a logical step to apply the Born (1926) rule for determining probabilities of different (physical) worlds occurring in that

with waves than if one assumes the existence of quasi-classical particles. But he would certainly not have abandoned his 'idealistic attitude.' Most certainly, Heisenberg was no proponent of the multiverse, so that this would have, anyway, been the more serious step in the context of my discussion.

stage, if one supposes that *each* of those realities must occur with the *same* observer with a probability of *one*, according to the deterministic Schrödinger equation. Especially instructive is the perspective taken by many-minds theorists (for more details see Chap. 4). Many-minds theorists will indeed see a plethora of 'perspectives' by different versions of the observer, but only *one* multiverse and no actual 'splitting.' This all will be associated with something that happens in consciousness, although in different ways in different versions of many-minds theories. Within Everett's original theory, this step already happens with the correlated stage, just that we do not realize this because *we* are entangled with only *one* of the relative states.

Hence conceptually, the last step might be seen as being almost identical with the second, a bit depending, perhaps, on which version of the multiverse interpretation will be adopted. It therefore depends on what kind of position is taken with respect to decoherence II and which multiverse interpretation one selects, whether this should be considered a second, a third or no cut at all.

What exactly would 'happen' if we stopped decoherence modelling before decoherence II, with entanglement (or decoherence I) being the last stage of the quantum analysis? Zurek (2002), in the above quotes, stated that this is tempting on the one hand, but would have a large disadvantage on the other hand. The disadvantage being that we would not be able to come up with a *realist* 'explanation' of how our quasi-classical reality emerges. I am not quite convinced that Zurek (2002) is right. What indeed happens if we, conceptually, drop the decoherence II stage is that we *openly admit* that we are not able to come up with a detailed 'story' as to what *physical processes* take place between the stage of quantum correlations (decoherence I), anyway *subjective* (see also Box 2.1, below), and our conscious perception of a singular, quasi-classical reality. *Perhaps none.* (And would we be any better off, in this regard, if we calculated the reduced density matrix? I doubt it. This is just important for empirical/practical purposes.) Perhaps decoherence I is indeed the end of a conceptual analysis within physics. This situation has two aspects. First, higher-level quantum correlations are fully sufficient to create quasi-classical realities (Zeh 2016a; Mensky 2010; Mensky 2000b); decoherence II is simply not required. Second, it is unclear *who* (which version of consciousness of the observer) will reside in which reality. The second aspect is actually serious, it is related to Everett's idea of relative states and will be analyzed in more detail in Chap. 4.

Box 2.1: Quantum correlations in a top-down view

It should again be pointed out here—it has only indirectly been mentioned so far, in connection with the above discussion of Heisenberg's cut—how close the position that could be (and actually will be) taken in this book, based on the presented analysis, is to Heisenberg's perspective on the role of the observer's consciousness (see also the discussion in Zeh 2013, 97–99): "(…) measurement selects a definite value for a measured observable 'from the totality of possibilities and limits the options for all subsequent measurements'" (Beller 1999, 67, partially referencing Heisenberg 1927, 74). "The source of this idea, as Heisenberg pointed out a few years later, in 1932, (…) was Fichte's philosophy of the self-limitation of the ego: 'The observation of nature by man shows here a close analogy to the individual act of perception which one can, like Fichte, accept as a process of the *Selbst-Beschränkung des Ich* (self-limitation of the ego)" (Beller 1999, 67, partially referencing a lecture held by Heisenberg at the Acadamy of Science in Saxony, 1932). Unlike Heisenberg's position, an idealistic version of the Copenhagen interpretation, the position taken in the current monograph is based on the *universal wave function* and it is hence opting for a *multiverse interpretation*, so that the self-limitation of the ego applies to *each version* of the observer, but the entirety of versions—looked at together—would not be self-limited. One might think of this perspective as linking Heisenberg's philosophy, Schrödinger's mathematics and Everett's interpretation. The consequences of this step will become clearer throughout the remaining chapters of the book. It makes sense, however, to present the basic position in a nutshell already here: Correlations have no direction, they do not imply any direction of causality. That is as true with quantum correlations as with any other correlations. Since the only difference between closed and open system perspectives is the existence of consciousness within the open system, all quantum correlations might be seen to *start* rather than end with the observers' consciousness. That means that I will consider the entanglement process *top down*, from consciousness to experienced reality. This is a purely subjectivist (or idealist) perspective, indeed close to a multiverse version of Heisenberg's singular-universe perspective. A purely subjectivist perspective is also briefly mentioned (but not pursued) in the monograph by Wallace (it is actually Savage's 1954 view) and implies that even the "half-life of uranium-235 is not an objective fact about the Universe but no more than some collective agreement among scientists" (Wallace 2012c, 138). I would like to push even further and speculate that in a *deterministic* universe, governed by the Schrödinger equation, the fact that a specific version of the observer measures the decay of a uranium-235 atom at a certain point in time might be rooted in certain decisions made by consciousness. Moreover, I would like to argue that Wallace is somewhat misleading when he continues: "(…) that there is a collective agreement as to the half-life is not in dispute, of course, what is in dispute is whether or not there is something objective about which they are agreeing" (Wallace 2012c, 138). I would like to instead argue that the collective agreement is *intersubjective* in a sense that whenever it is reached, everyone within a certain '*minds cluster'* (see Chap. 4) will get empirical results that are in tune with the agreement. Specifically, whenever the scientific community within a cluster has reached an agreement and published the half-life of uranium-235, nobody will ever get findings falsifying that result.

"What Is the Correct Interpretation of Quantum Mechanics?"

In the popular comedy series "The Big Bang Theory,"[33] Dr. Sheldon Cooper asks his (only) prospective student Howard Wolowitz exactly the question heading this section. And since Howard wants to be accepted as Sheldon's student, he answers: "As every interpretation gives exactly the same answer to every measurement they are all equally correct. However, I know you believe in the many-worlds interpretation, so I'll say that."

So, is it just a matter of *belief* which interpretation to accept? I feel that I am able to say a bit more after revisiting the evidence that was presented in this chapter so far. In principle, the multiverse interpretation is the *most parsimonious* interpretation of quantum mechanics in light of the measurement problem since it normally does not require changing anything in the wave equation; as will be shown in Chap. 4, however, this statement is not correct for *all* versions of that interpretation, but it will be correct for the version I am going to propose. Moreover, the multiverse interpretation still requires the Born (1926) rule or decoherence II (the reduced density matrix) as auxiliary equation or stage, respectively, to make predictions about the relative frequency of outcomes with repeated measurements within one reality (for more details see also Chap. 4). With respect to the Born rule, there is, anyway, a standoff between different interpretations of quantum mechanics; they all have to postulate it, without any good *conceptual* justification (despite its straightforward mathematical calculation via mod-squared amplitudes and its excellent empirical support).[34] It also became clear that the decoherence principle per se is not the 'white knight' preserving quantum mechanics from the necessity to solve its measurement problem. It rather delivers very important insights, and it actually helps with better understanding the measurement problem. Anyway, the universality of the wave equation is not questioned, no changes are required under *any* conditions, no postulates such as collapse/reduction are needed in the multiverse.

But it might then be argued that the price to be paid for the higher parsimony of the multiverse interpretation (in most of its versions) is (too) high. We are not able, some will say, to come up with an account of the appearance of our *singular* quasi-classical world—as perceived in consciousness by different individuals. But

[33]Season 8, episode 2, "The Junior Professor Solution," CBS, Monday, September 22, 2014.

[34]And unlike the Oxford school, I do not think that conceptually, it makes sense to derive the Born rule from decision-theoretic principles applied to the Schrödinger equation (see Chap. 4).

this is incorrect. In fact, Everett's idea of *relative states* (and nothing else) is honest and simple. It, however, moves our area of explanation (partially) away from physics (at least from the viewpoint of the author of this book).

Box 2.2: Is the multiverse interpretation saving quantum mechanics from the 'locality problem'?

Some authors claim that there is another seeming advantage of the multiverse interpretation over other interpretations of quantum mechanics. Quantum mechanics has been criticized, e.g. by Einstein, for its irritating non-locality, its 'spooky action at a distance' problem (spukhafte Fernwirkung) (see, e.g., Einstein et al. 1935 (EPR); Bell 1964). But some authors consider the multiverse interpretation to be able to deal with that problem, to make quantum mechanics 'local.' According to Squires' (1994) understanding of the standard interpretation of quantum mechanics, e.g., "(…) [a] local description of reality which permits us to talk of objects which are spatially isolated from each other does not exist. We have here reached the ultimate 'silliness' of the quantum world. The lack of locality in quantum theory was seen in the very early days by Schrödinger as its prime feature (…). The only way it can be avoided is to adopt the many-worlds interpretation." (105) Another prominent author claiming that the multiverse interpretation is 'local' is David Deutsch (Deutsch and Hayden 2000; Deutsch 2011). However, the mathematical proofs that these authors present are based on Heisenberg's picture assuming the existence of 'particles,' thus leave the universality of the wave function, otherwise assumed in this monograph. Bacciagaluppi (2001) uses the wave function and also proves locality in the multiverse, albeit in a relativistic picture, i.e., in Minkowski space time (Minkowski [1908] 1952). The advantage of 'locality' of the multiverse interpretation is also mentioned by Lockwood (1991, 217). In an online forum, Domino Valdano (2014) even gives a nice figural description as to how locality can be preserved in the multiverse; he directly addresses the problem of entanglement between distant particles where two distant observers are able to measure correlated results: "(…) in the many worlds interpretation, even (…) large classical observers are initially in giant superpositions (albeit decoherent ones) of having measured both results (not one or the other). It's not until they come into causal contact with each other that the branches of the multiverse which involve non-correlated results fully cancel out with each other (…). In other words, the information about which result they got travels through a classical channel at less than the speed of light." The most *consequent* multiverse proponent in terms of staying with pure wave mechanics of the Schrödinger equation, however, is Zeh. In his 2016a paper on the (hi)story of particles and waves, he clearly argues *against any locality* in connection with the wave equation. And he traces back the misunderstandings of the concept to its inventor: "Schrödinger was convinced of a reality in space and time, and so he originally hoped, (…) to describe the electron as a spatial *field*. Therefore, he first restricted himself with great success to single-particle problems (quantized mass points, whose configuration space is isomorphic to space). (…) In spite of its shortcomings, three-dimensional wave mechanics still dominates large parts of most textbooks because of its success in correctly and simply describing many important single-particle aspects, such as atomic energy spectra and scattering probabilities" (Zeh 2016a, 199). The description of more complex problems, however, requires much higher dimensionality, it requires wave functions in (higher-dimensional) configuration space (Zeh 2016a). For me this implies that the complex question of 'locality' within the multiverse interpretation is not to be decided in this book. This would require a much more intensive treatment of the matter. The good news is that the theory and applications to be developed here are completely independent of the answer to this question. The bad news is that 'locality' cannot unambiguously be used as an argument in favor of the multiverse interpretation of quantum mechanics.

A *disproval* of the multiverse interpretation is on the agenda of the philosophy-of-science paper by Emily Adlam (2014). Adlam uses internal consistency arguments against the multiverse interpretation. Adlam's argument has two parts. One part is specifically addressed to disprove the 'proof of the Born rule' (the probability rule of quantum mechanics) within the Everett interpretation, suggested by David Wallace. This part contains one more (of many) arguments that have been made in a discussion on how (not) to make sense of the Born rule in the multiverse. This part of the argument will not be addressed, here. Probabilities and the Born rule in the multiverse are, however, important topics that have already been mentioned and will be dealt with in more detail in Chap. 4.

The other part of the argument is more generally crafted against all kinds of Everettian approaches suggesting that in the multiverse, consciousness is residing with many versions of a person (after a measurement); this includes, e.g., Wallace's approach as well as Mensky's, Deutsch's and my own. Adlam wants to show that this type of a multiverse does not qualify as a *scientific theory* for lack of internal consistency. Specifically, Adlam demonstrates, using *Bayesian-type* arguments (an empiricist approach using sophisticated probability calculus), that an individual agent will not be able to increase his *credence* in (Everettian) quantum theory by appropriate experimentation if Everettian quantum mechanics is true. In a way, she uses the fact that different versions of an agent reside in different parts of the multiverse after an experiment against Everettian quantum theory. This agent can just learn *where* (in which part of the multiverse) he is and not more. Somehow, an outcome of an experiment can only be used for one purpose, and that purpose is dictated in the multiverse by the generation of *self-locating information*. So, in narrow terms, scientific experiments do not work if we are in an Everettian universe, and that includes experiments on Everettian quantum mechanics.

Deutsch (2016) does not specifically deal with Adlam's set of arguments but addresses all kinds of Bayesian arguments that might be used against the testability of Everettian quantum theory and is hence *relevant* for Adlam's type of argument. Deutsch's proof extends and sharpens arguments taken from Karl Popper's (1959) critical rationalism. Deutsch (2016) shows, that (among other requirements) if "scientific theories [are considered] as conjectural and explanatory and rooted in problems (rather than being positivistic (…))," Everettian quantum theory can be tested. Deutsch leaves the single-agent paradigm of Bayesianism and exploits the fact that scientific theories are crafted for a purpose (explanation). Let me add here that (closely related to Deutsch's set of arguments) scientific theories are tested within a multi-agent, multi-round (repetitions!) scientific process including trial-and-error processes and intensive scientific discussions. Within such a process, theories are developed and supported, falsified, discarded etc. A Bayesian argument (including Adlam's) is an undue simplification of that scientific process. Hence there are good reasons not to buy into Adlam's disproval of the possibility to confirm Everettian quantum theory. Let me finally add that part of any scientific process is interdisciplinary research and *interdisciplinary evidence*. Such evidence exists for the multiverse interpretation as will be shown in the next chapter.

Quite promising in terms of *support* for the multiverse interpretation is the proof by Frauchinger and Renner (2016) of the self-inconsistency of single-world interpretations of quantum mechanics. In their paper, the authors construct a dilemma for single-universe quantum theory. Specifically, they "propose a gedankenexperiment where quantum theory is applied to model an experimenter who herself uses quantum theory." (1) In their mathematical proof[35] they show that there "cannot exist a physical theory T that has all of the following properties:

(QT) Compliance with quantum theory: T forbids all measurement results that are forbidden by standard quantum theory (…).
(SW) Single-world: T rules out the occurrence of more than one single outcome if an experimenter measures a system once.
(SC) Self-consistency: T's statements about measurement outcomes are logically consistent (even if they are obtained by considering the perspectives of different experimenters)." (2)

A central part of their proof involves multiple observers, partially residing within different *isolated quantum systems* (Frauchinger and Renner 2016, 11–16). However, in their proof, experimenters must communicate with each other and an open question, at least from my perspective, is *how* they might do so if they are residing in isolated systems. So, whereas I find their inconsistency proof quite stimulating, it still awaits an intensive discussion by the scientific community.

Another support for the multiverse interpretation is offered by the fact that seemingly paradoxical results of quantum thought experiments as well as actual

[35]In the following I am only presenting their set of propositions, and also in an informal version. For the formal version and proof see Frauchinger and Renner (2016).

experiments may arise *only* when a singular-reality framework is used for the explanation of the (to be) observed effects. Using a multiverse perspective, many such paradoxes disappear, and the results of such experiments indeed appear reasonable (Vaidman 1994, for two striking examples for this).

After all, the evidence in favor of the multiverse interpretation within physics (and closely related discussions in the philosophy of science) is mixed to positive. The most important advantage of the multiverse interpretation over other interpretations of quantum mechanics is the relatively higher parsimony (with most of its versions), its (almost) sole reliance on Schrödinger's wave equation (see Chap. 4). It normally does not comprise any severe, artificial postulates implementing changes to it (such as a wave function collapse)[36]; only the Born (1926) rule has to be assumed as a practical tool or decoherence II carried out to generate empirical predictions—as is the case with all other interpretations of quantum mechanics. Moreover, it cannot simply be rejected by testability arguments in the spirit of Bayesianism, and there are perhaps even ways to show internal inconsistency of singular-world versions of quantum mechanics—those approaches being so recent that they are still awaiting an intensive scientific discussion.[37] But even if the evidence for the multiverse interpretation would be clearer at this point, it would still be unclear whether this would be sufficient to actually encourage a shift of paradigm towards a multiverse view (both within and outside physics). Paradigm shifts are difficult; they have to overcome various hurdles (Kuhn 1996 [1962]). So, there are many good reasons as to why Chap. 3 is presenting an interdisciplinary search of evidence for the multiverse.

[36]An exception are those multiverse versions where infinite minds with relative frequencies resembling the probabilities given by the Born (1926) rule are postulated. This would require severe changes in the quantum formalism and will be critically discussed for that reason in Chap. 4.

[37]Not mentioned so far, there might still be another possibility to conduct a critical test of the multiverse interpretation. David Deutsch believes that such a critical test might be possible with the help of quantum computers in the future (Deutsch 1985). However, given the fact that the stage of development of quantum computers is still in its infancy, this might not be a possibility one could, in a narrow sense, wait for in a lifetime.

Chapter 3
Interdisciplinary Evidence for the Multiverse, Including a Detailed Analysis of What Time Is

Can Research from Outside Physics Help Solve the Interpretation Problem of Quantum Mechanics?

The last chapter showed that the evidence within physics and philosophy of science suggesting some version of the multiverse being the 'correct' interpretation of quantum mechanics is mixed to positive. Clearly, nobody is *forced* to believe in this interpretation at this point since direct experimental tests comparing different interpretations within physics have not been carried out so far, are indeed viewed as impossible (at least in the present). Given this impasse, could an *interdisciplinary perspective* be helpful? An endeavor that might be understood as an interdisciplinary, 'presumptive-evidence' proof for the existence of the multiverse will be undertaken in this chapter. The basic idea as well as the ingredients of the 'proof' are depicted in Fig. 3.1.[1]

The logic of reasoning is as follows. There are several interpretations of quantum mechanics, and any of them might be adequate. There are also several theories of consciousness (in light of neuroscience findings), and many of them appear to be reasonable. Then there are several philosophical positions on the existence/non-existence of free will. Which one is appropriate? Finally, how could one explain seemingly 'strange' phenomena such as predictive physiological anticipation (see below)? Within each of the disciplines those questions are undecided, hence there are several, quite different 'pieces' that the respective discipline might contribute to the overall 'puzzle.' Trying to merge all these disciplinary perspectives to one picture, i.e., trying to fit the disciplinary pieces to one overall puzzle, however, only one version of theorizing out of each of the respective fields qualifies as an appropriate part of the overall picture. Everything only fits in exactly one form, and in turn a selection of one out of the several possible alternatives within

[1]This chapter partially builds up upon Schade (2015), but largely modifies and extends this early approach, especially with respect to time.

© Springer Nature Switzerland AG 2018

C. D. Schade, *Free Will and Consciousness in the Multiverse*,
https://doi.org/10.1007/978-3-030-03583-9_3

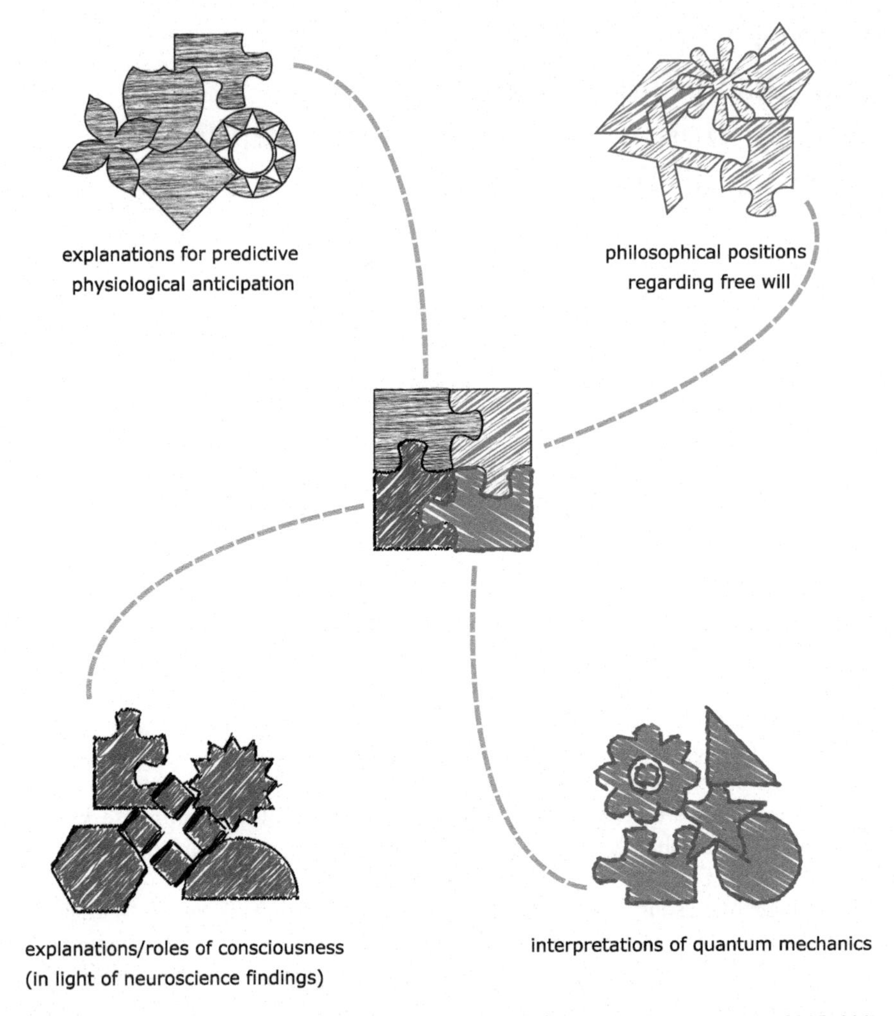

Fig. 3.1 An interdisciplinary treatment of the measurement problem. *Source* Schade (2015, 331)

each of the contributing disciplines is feasible. Or in other words, all those questions can simultaneously be addressed via an interdisciplinary effort. This includes, and that is the main point here, the interpretation problem of quantum mechanics, the free will problem, and the consciousness problem. It also allows for a physical basis for phenomena such as predictive physiological anticipation.

The last chapter discussed different interpretations of quantum mechanics (especially the Copenhagen interpretation—as an umbrella term for a couple of single-world interpretations of quantum theory including von Neumann's reduction postulate—and the multiverse interpretation) as well as the principle of decoherence. This will not be repeated, here. The only ingredient that is needed here is the result that the multiverse is one of the *theoretically sound* interpretations existing of quantum mechanics (actually one of two mainstream interpretations). In the following, the phenomenon of physiological anticipation will be presented first. It will be shown that there is *no explanation* for this phenomenon *outside the multiverse*, but that some versions of the multiverse interpretation might provide a physical basis for its existence. A detailed account of the different philosophical positions on free will is to be postponed to Chap. 6. Here it will be sufficient to see that an *actual free will* (in the sense of an existence of alternative possibilities and some influence of the individual on what will be experienced to what extent) is *only* possible in the multiverse, and that the possession of an actual free will would help to make sense of the existence of consciousness, of *qualia*, that it would give our conscious experience a *meaning*.

After presenting the analysis of physiological anticipation within the framework of the multiverse, the chapter will turn to (additional) evidence for *parallel times*. This turns out to be of vital importance since the solution to the puzzle of 'anticipatory' responses proposed is based upon this notion. In this section, the chapter returns to physics, partially building up upon the previous chapter, but also extending the scope of the theory. Starting with, e.g., implications from *special relativity*, from *block universe* physics (e.g., Minkowski 1952 [1908]; Petkov 2005; Silberstein et al. 2018) and from the Wheeler-DeWitt equation (DeWitt 1967), it will then critically reflect upon potential lessons that may or may not have to be taken from a decoherence-based theory of the time arrow. After a short section on theories of consciousness and potential meanings of qualia, I will then be able to reinterpret the Libet evidence and 'free' the free-will debate of the heavy burden it has carried since the 1970s. At the end of the chapter I will summarize how all those pieces now fit together.

A Theoretical Framework for Predictive Physiological Anticipation

Are peoples' bodies able to anticipate future developments? Mossbridge et al. (2012) present surprising evidence for this. And not just from one study. In their meta-study, actually based on 26 reports, they present robust results pertaining to all kinds of body reactions such as "electrodermal activity, heart rate, blood volume, pupil dilation, electroencephalographic activity, blood oxygenation level dependent (BOLD) activity" (1). The bodies either react to randomly ordered arousing versus non-arousing tasks; i.e., to the fact that *something exciting is going to happen* but without getting any prior access to this later development by 'classical' means. Or they react to guessing tasks with correct/incorrect feedback, also *before* any feedback was provided. The statistical significance over all those 26 studies is so high that 87 unpublished contrary reports would be needed to reduce this significance to chance.[2] The evidence is strong enough to look for possible explanations of such a phenomenon. Critics of such phenomena might say that something like anticipation of the future, no matter whether the bodies or anything else are supposed to anticipate it, simply cannot exist despite all that evidence because there is no physical basis for this. But is that true? What is the situation in physics regarding the explanation of such a phenomenon? Is it at least possible to provide a *general framework* for this?

The question one wants to ask here is whether there might be ways of thinking of time as something that does not just flow in the way we normally assume, from the past to the present, and then from the present to the future? One way to go would be to time reverse physical laws, especially those from quantum mechanics. Basic physical equations, including the Schrödinger equation, *are* fully reversible, and even quantum measurement might be treated in a *time-symmetrized* way (see Aharanov et al. 1964; see also Price and Wharton 2015; Elitzur et al. 2016; Sheehan 2006; Vaidman 2012).[3,4] But there are nevertheless two problems with this. One problem, often mentioned in this regard, is perhaps not so substantial, the fact that the second law of thermodynamics appears *not* to be reversible. This problem will be dealt with in some detail below, within Box 3.1 towards the end of this chapter.

The other problem with time reversing physical laws is more severe. This approach actually collapses for a logical reason, i.e., for a reason that has been most clearly articulated in the so-called *grandfather paradox* in the literature on *time travel*.[5] Let me suppose that you are travelling back in time. You arrive at a time where your grandfather was young, actually a point in time where your grandfather

---

[2]It is probably less surprising than in the macro sphere that such phenomena are observed on the quantum level. 'Time-backwards' effects are one explanation that has been put forward for the results of so-called quantum-eraser experiments (see, e.g., Herzog et al. 1995). But other interpretations for those findings have also been proposed.

[3]A detailed discussion of all these approaches is beyond the scope of this book. Regarding the famous but also problematic approach by Aharanov et al. (1964), the following discussion of this single-universe interpretation of quantum mechanics in Vaidman (2012, 583) might suffice: "There is a certain difference between the single world described by quantum mechanics with collapses at each measurement, and the single world which emerges with the backwards evolving quantum state of Aharonov. While the former, at each moment in time, is defined by the results of measurements in the past, the latter is defined in addition by the results of a complete set of measurements in the future." And at a later point in the paper, Vaidman states: "I find Aharonov's proposal very problematic. It does remove action at a distance and randomness from basic physical interactions, two of the main difficulties with the collapse postulate. But it still has the third: it is not well defined. The backwards evolving quantum state needs to be tailored in such a way that all measurements will have a definite result, but what is the definition of a measurement?" (2012, 587).

[4]Silberstein et al. (2018) discriminate between two types of so-called retrocausal theories: "So looking at retrocausal accounts more generally, it seems that there are two basic ways to go, one we call "time-evolved" or "retro-time-evolved" and the other we call global (4D). The former focuses on positing (relatively) new dynamical mechanisms to underwrite retrocausation and the latter takes a more global, adynamical approach" (195).

[5]I am presenting this problem here in a slightly less violent version than usual.

has not even met your grandmother, yet. You now interact with your grandfather at the very day where he would have met your grandmother, and he does never meet her. Since your father will never be born, you should not exist. In a singular reality, you have produced a *logical inconsistency*.

Sure enough, we are not up to explaining time travel but the anticipation of future events. But the latter potentially leads to the same type of logical paradox than the prior. The paradox appears in a singular universe if your body's anticipation of the future changes that future and is then inconsistent with the new future. This can be clarified via the use of a simple example. Let me suppose that your body anticipates a negative future event and simply reacts to that prospect. Specifically, let me suppose you walk down a city road, northbound. Your body correctly anticipates that something very negative will happen to you if you continue to walk that way with continuous speed. Your legs begin to tremble and you stop, then you even sit down in a nearby café and drink a cappuccino (You have no idea why your body did that). When you later continue, nothing negative happens to you, since the 'window of opportunity' within which the event would have taken place at the respective location has passed (Say, a car backed up quickly without the driver looking backwards, but you were not there). Thus, the reaction of your body has changed the future. But since nothing happens, the prior reaction of your body has now been unjustified, i.e., it turns out to be inconsistent with this new reality.

How would a theory have to be structured to account for this? There is indeed a more radical way than time reversing physical laws to allow for the anticipation of future events. And this is to totally give up the idea that time *flows*—no matter in what direction—and to allow for a *certain type* of parallel realities. As a matter of fact, this is David Deutsch's solution to the grandfather problem (Deutsch 1991; Deutsch and Lockwood 1994). The idea is challenging: There is *no* logical inconsistency if a time traveler, after travelling to the 'past,' interacting with his grandfather, causing him *not* to meet the traveler's grandmother and then returning to the 'present' resides in a different reality, parallel to the reality where he had changed the 'past.' In the 'past' of the reality where the traveler now resides, the grandfather *has* met the grandmother. Then there also is a reality where the grandfather did not meet the grandmother and the time traveler does not live in the 'presence.'[6]

Similarly, in the above example where you walked down a city road there might be one reality where you have somehow managed—despite your trembling legs—to continue northbound with constant speed and something very negative happened to you (The car, say, backed up, and you are hit by it). But this, again, is not where 'you' are now.[7] Instead, you now reside in a reality where nothing negative took place. It is easy to see that the only consistent framework for

[6]The wording assumes some kind of consistent histories, and this actually helps discussing the situation. In fact, not even this assumption has to be made.

[7]Sure enough, in the multiverse you will be residing in all those realities, but consciousness may have put its main emphasis on a different reality; see Chaps. 4 and 6, for this type of reasoning.

predictive physiological anticipation is the multiverse. Your body anticipated a development in a *different* reality. Its reaction was consistent with what was going to happen in that reality. Since you slowed down and had a cappuccino, this reality has not been reached (you changed reality!), and you are now safe.[8] Such type of anticipatory reaction is only possible in the multiverse, and only in special types of it, too.

How would those types of the multiverse have to look like? In his monograph *The Fabric of Reality* Deutsch (1997, 278) states: "Other times are just special cases of other universes." In the eleventh chapter of his book, he develops time as a fourth dimension (additional to the three of space) and fully abandons the idea of a *flow* of time. For Deutsch, consciousness is only taking 'snapshots'[9] of certain versions of reality at certain points in time. Mensky (2010; 2007b) proposes a similar view.

But Are Times Really Parallel?

How can Deutsch's and Mensky's approaches be justified? Do they in fact naturally align with the multiverse interpretation, or perhaps: some version of it? And what do other physical theories have to say on this? Furthermore, would everyone agree with Lockwood (1991, 13), stating that "(…) [t]emporal flow has no place within the physicist's world-view, so we must consign it to the mind?" Aren't some authors even claiming that decoherence, already discussed in the last chapter, 'produces' something like a time arrow (e.g., Zeh 1999; Wallace 2012a, b), no matter whether a Copenhagen or a multiverse view is assumed? And is this in contradiction with the above explanation proposed for predictive physiological anticipation or not?

A so-called *non-presentist'* view of time (consistent with the so-called B-theory of time; Gale 1966; McTaggart 1908) where all times are assumed to exist in parallel might be, simplifying, derived from two different lines of reasoning: (a) block-universe physics, especially pronounced in its non-dynamic, 'frozen-universe' version (Silberstein et al. 2018) and (b) quantum gravity, most striking in terms of the results of the Wheeler-DeWitt (DeWitt 1967) equation and Barbour's (1999) as well as Deutsch's (1997) interpretations.

(a) The block universe is a direct consequence of Einstein's theory of *special relativity* with its challenging problem of *simultaneity*. Indeed, Minkowski

[8]The last footnote fully applies here, too.

[9]Note that whenever I use the term 'snapshot,' I am aware of the fact that these are 'interesting snapshots.' Unlike the pictures taken by a camera in a singular, classical reality, the kinds of snapshots consciousness is taking in the quantum multiverse are not objective.

proposed the *block universe view* with time as a *permanent* fourth dimension as a reaction to special relativity already in 1908 (1952, 75). And because multiple times coexist, multiple spaces do as well:

> We should then have in the world no longer space, but an infinite number of spaces, analogously as there are in three-dimensional space an infinite number of planes. Three-dimensional geometry becomes a chapter in four-dimensional physics. (*Ibid.*)

This early conjecture by Minkowski has been substantiated by more recent research in theoretical physics. According to Petkov (2005), the parallel existence of different times (and spaces) *is required* from special relativity theory. The most recent as well as most developed approach in the field of block universe physics is the *relational-blockworld* theory, spelled out in detail in the recent "Beyond the Dynamical Universe"-book (Silberstein et al. 2018).[10] Simplifying, those authors state that *everything is given.* We are living in a 'frozen' universe and only subjectively 'experience' dynamics, i.e., a flow of time. The relational blockworld is a *singular universe.* On the one hand, this fact rules out the possibility of using this theory to explain physiological anticipation in the way specified above. On the other hand, there would simply be *no need for* an *actual anticipation* of one out of many possibilities, i.e., a *prediction* in a narrow sense; consciousness would have to 'only' *read out* a singular future (what might certainly still be difficult to impossible in practice). And time travel would never lead to any changes in the past having an impact on the future—because no changes are possible in a 'frozen' universe, neither in the past, present or future (see for the authors' own, slightly different discussion Silberstein et al. 2018, 115–121).

So, one might ask here the more general question as to how plausible the theory by Silberstein et al. (2018) is, although a comprehensive account and criticism of a densely written 400-pager clearly is beyond the scope of the current book. The formal development of their theory is generally impressive.[11] But there are potential conceptual problems. Even though the authors propose an adynamical framework, I am missing an open and detailed discussion of the question as to whether their approach is in fact a *hidden-variable theory.*[12] They describe their theory as replacing anything like this by so-called *adynamical global constraints*;[13] and they also claim not to need anything such as measurements that

[10]This book also contains an extensive review of the literature in this and related field(s).

[11]Just a small example is the relational-blockworld analysis of the double-slit experiment (Silberstein et al. 2018, 209–213).

[12]The most important proponents of a hidden-variables theory, de Broglie and Bohm, are mentioned once in the introduction of the Silberstein et al. (2018) book ("overture for ants") and they are not referenced once in that book. Entries such as 'hidden variable' or 'third variable' are also missing in their index. However, hidden variable approaches are nevertheless touched in some parts of their book, e.g. on page 149.

[13]These can be seen as the constraints that have been used when 'constructing' the 'frozen' universe. This concept is at the core of their theory and appears in several parts of their book (see, e.g., the discussion in Silberstein et al. 2018, 149).

are jointly determined by past and future measurements to reach a singular reality, the approach used within certain retrocausal approaches.[14] But what does it exactly *mean* to have such constraints in the context of measurement? This question is inevitable, from my point of view, given the multiplicity of possible outcomes in the measurement of quantum systems together with the result by Silberstein et al. (2018) that there is only *one* 'frozen' universe; in other words, *how do the authors address the measurement problem* (see Chap. 2 of the current book)?

Do they implicitly assume some *pre-randomization* according to the Born rule (i.e., by the creator of the relational blockworld) before that universe is 'frozen;' and is that pre-randomization (or the result of it) becoming part of the adynamical global constraints?[15] The authors deny that the measurement problem is even relevant to them: "(…) the measurement problem is a non-starter for us. When a (…) [quantum-mechanics] interpretation assumes the wave function is an epistemological tool rather than an ontological entity, that interpretation is called 'psi-epistemic'. In (…) [the relational blockworld], the wave function in configuration space isn't even used, so (…) [the relational blockworld] is trivially psi-epistemic" (Silberstein et al. 2018, 160). What is also puzzling is another central piece of their theory, the somewhat obscure *spacetimesource elements*. In the context of measurement, the authors describe them in the following way: "We will argue that the outcomes observed in quantum experiments are the result of 'spaciotemporal ontological contextuality,' a 4D contextuality that includes the experimental setup from initiation to termination. Thus, our fundamental ontology is 4D spacetimesource elements (…), (…) for now note that they include properties that would normally be considered classical and properties that would normally be considered quantum" (Silberstein et al. 2018, 137). For me this is hard to follow since I see this definition as partially tautological. The authors are somehow defining spacetimesource elements via spaciotemporal ontological contextuality. It is, anyway, hard to understand what this contextuality is supposed to grasp, and what it means to include properties that are classical and quantum. Unfortunately, the entire concept does not become any clearer in later parts of the book, but more questions are added: "Spacetimesource elements constitute our fundamental ontology and, since they don't represent 3D entities moving in space as a function of time, they are best viewed as being *of* spacetime, not *in* spacetime" (Silberstein 2018, 153).

Since for me, the wave equation is at the core of quantum mechanics and has an 'ontological status' (i.e., is actually saying something meaningful about reality), their line of reasoning couldn't be farther away from mine. There are a couple of

[14]This idea was first developed by Aharanov et al. (1964).

[15]Such an interpretation is indirectly suggested when the authors, using Feynman path integrals for their analysis, state: "(…) the probability that the Feynman path integral assigns to a particular experimental arrangement to include a specific experimental outcome refers to the frequency of occurrence of these 'experimental regions in spacetime' per the God's-eye view" (Silberstein et al. 2018, 153).

generally deep and complex chapters on free will and experienced time in the authors' relational blockworld. However, those chapters remind me a lot of 'compatibilist-type' arguments, known also from other authors in the free will and responsibility debate (see, for a discussion of compatibilist' arguments, Chaps. 7 and 8 of the current book): "As regards free will, if the reality of the future (however it got there) is sufficient to make one conclude that free will is an illusion or some such, then so be it. (…) So as many have pointed out, it is one thing to say that I can't *change* the future and quite another to say that I can't *affect* the future" (Silberstein et al. 2018, 383; italics in the original). For me, free will presupposes that *alternative possibilities* exist and that the individual has some influence on what will be experienced to what extent (see Chaps. 4 and 6). Summing up, the relational blockworld theory by Silberstein et al. (2018), one of the most developed blockworld concepts so far, replaces one clearly defined entity, the universal wave function, with two less clearly defined ones: adynamical global constraints and spacetimesource elements. The latter are actually obscure and of unclear nature, physically. Furthermore, the consequences of the theory, e.g., for free will, are not encouraging. Therefore, I am not pursuing this approach any further in this book.

(b) Another radical view on time can be derived from *quantum gravity* (canonical theory). Specifically, the Wheeler-DeWitt equation, formally combining the basic structure of quantum mechanics and *general relativity* (DeWitt 1967), *does not contain time*.[16] The fact that time disappears from the differential equation has been labeled the 'problem of time' (see, for a detailed analysis, e.g., Anderson 2012). The Wheeler-DeWitt equation as well as the 'problem of time' might be reflected upon critically: "Of course, one needn't accept this interpretation of the Wheeler-DeWitt equation or even accept the canonical quantization method itself as there are other approaches to (…) [quantum gravity]"[17] (Silberstein et al. 2018, 34). Barbour (1999), one of the firmest proponents of the Wheeler-DeWitt equation and its interpretation in the form of the 'problem of time'[18] also admits: "Ironically, DeWitt himself thinks that it is probably not the right way to go about things, and he generally refers to it as 'that damned equation.'" (39) But then Barbour continues: "(…) However, there is no doubt that the equation reflects and unifies deep properties of both quantum theory and general relativity. (…) For a long time, physicists shied away in distrust from its apparently timeless nature, but during the last

[16]This equation was first called Einstein-Schrödinger equation.

[17]Other approaches to quantum gravity will not be dealt with in this monograph.

[18]Barbour (1999) himself builds up on the ideas by Ernst Mach (1883). Consistency with the multiverse interpretation is not denied, but the relationship is not explicitly developed, perhaps for personal reasons: "*Can we really believe in many worlds?* The evidence for them is strong. The history of science shows that physicists have tended to be wrong when they have not believed counterintuitive results of good theories. However, despite strong intellectual acceptance of many worlds, I live my life as if it were unique. You might call me a somewhat apologetic 'many-worlder'" (Barbour 1999, 324).

fifteen years or so [i.e., since the mid 80s of the last century, if one takes the time of publication into account] a small but growing number of physicists, myself included, have begun to entertain the idea that time truly does not exist" (Barbour 1999, 39). Andersen actually discusses a few approaches to deal with the problem of time. He calls the approach favored by Barbour (1999) as well as me the *'tempus nihil est'* (i.e., 'time does not exist') approach (see, for a couple of theoretical approaches to model 'tempus nihil est,' Anderson 2012, 9–11).
A major strength of the Wheeler-DeWitt approach is that it builds up upon the Schrödinger wave equation. By that means, the Wheeler-DeWitt equation is a multiverse equation, in the same way as the original wave function. It describes the entire universe as a quantum multiverse. But what to do with the fact that time is missing as a variable in the differential equation? Is there anything possible beyond stating that time does not exist? From the perspective of the current monograph, two approaches might be taken to deal with the problem of time, and both have their merits. One is Deutsch's above suggestion that times are parallel and just "special cases of other universes" (Deutsch 1997, 278). This is just a more sophisticated way of stating 'tempus nihil est.'
The other approach tries to *rescue* our usual notion of a flow of time somehow. This implies 'reconstructing' it or making it an *emergent* phenomenon. An early approach in this vein is the approach by Page and Wootters (1983). In their theoretical development, they show that entanglement (i.e., decoherence I) may lead to an *evolving* universe from the perspective of *internal observers* even though *objectively*, the universe might be a *static system*. Quite recently, Moreva et al. (2014) present confirmatory experimental evidence for Page and Wootters' (1983) theory. Moreva et al. (2014) construct a clock by smartly defining a movement as time: "(…) the clock photon is a true (albeit extremely simple) clock: its polarization rotation is proportional to the time it spends crossing the plates" (Moreva et al. 2014, 052123). Specifically, the authors "show how a static, entangled state of two photons can be seen as evolving by an observer that uses one of the two photons as a clock to gauge the time-evolution of the other photon. However, an external observer can show that the global entangled state does not evolve" (Moreva et al. 2014, 052123). For me, this clearly implies that the flow of time is *subjective*.
Being fully aware of this difference between internal and external observers, some authors nevertheless tentatively claim to be able to construct something like an 'actual time arrow,' at least according to my reading of their publications, out of irreversibility (see also Chap. 5). In his monograph *The Physical Basis of the Direction of Time* Zeh (1999), e.g., analyzes, among other theories, the second law of thermodynamics (37–82) (see Box 3.1, for a more detailed discussion of the implications of that theory for the concept of time) and the 'quantum mechanical arrow of time' (83–130); and he sees a formal analogy between the two, as Mensky does (for this reasoning see Mensky 2000b, Chap. 2, who, however, does not propose decoherence as *creating* an 'actual,' an emergent flow of time; see below). The 'quantum mechanical arrow of time'

is based on the principle of decoherence; so, in a way it is similar in spirit to the entanglement argument by Page and Wootters (1983), but, unlike in Page and Wootters (1983), seen as establishing something slightly 'more objective.' A similar type of reasoning is part of the *decoherent histories* theory within the realist' interpretation of the multiverse by Wallace (e.g., 2012a, c; see the discussion below). As with Zeh (1999), Wallace's main argument is *irreversibility*. Measured systems interact with the environment. This interaction is especially *effective* in the macro sphere. Because those interactions lead to larger and larger entanglement with their environment, such a process can *practically not be reversed.*[19]

For the sake of supporting the search for interdisciplinary evidence for the multiverse in the current chapter, one must ask whether the sketched, decoherence-based irreversibility is sufficient for establishing an *actual*, an objective flow of time and, thus, could become a problem for the explanation of predictive physiological anticipation. At the core of the discussion must be the question as to how *subjective* or *objective* is the flow of time. Only if emergent time where an *objective time*, effectively excluding consciousness from 'traveling' between different times, a problem for the principle possibility of predictive physiological anticipation (notwithstanding the necessity to clarify the exact mechanism in future research) would arise. In the following, however, I am offering a set of arguments as to why emergent time should be seen as *subjective* (and could hence not become a problem for the principle possibility of physiological anticipation).

(a) One should be doubtful that some practically, not even principally, irreversible process (entanglement/decoherence) is able to *'produce' objective time.* This reminds me a bit of the famous tale of Baron Münchhausen, a fictional German baron (rooted in an actual person, however), who, in one of the stories, pulled himself and his horse out of a mire by his own hair (Raspe 1785). So-called 'bootstrapping' methods have been successfully developed in some fields such as statistics (there a method of resampling exploiting the information in a sample; Efron 1979), but I doubt that bootstrapping is feasible with time in quantum mechanics.
(b) My next argument would be that *entanglement* (or decoherence) is anyway *subjective*, not only with respect to time, based on the top-down view proposed and defended earlier in this book (see my detailed discussion in Box 2.1 above). Since I see consciousness of the observer as the starting point for quantum correlations, I also see it as the starting point for subjective time. As already mentioned, Mensky also argues that the nature of irreversibility due to decoherence is only *subjective* in the framework of the multiverse (Mensky 2010, 121).
(c) One should look a bit deeper into the theorizing of the currently most prominent author within the paradigm of the decoherent histories framework (Wallace 2012a,

[19]Note that this reasoning is independent of the differentiation of decoherence into two stages that was introduced above but already applies to decoherence I or entanglement.

c). What is Wallace's position, here? Wallace, unfortunately, makes it rather hard for the reader to find out. Within an introductory discussion, however, Wallace states[20]: "I also avoid, in large part, two terms which at one point were common in the literature: 'many-worlds' theories and 'many-minds' theories. Both carry connotations I wish to avoid: in the one case, that the many worlds are somehow *fundamental* parts of the theory (…); in the other, that somehow a detailed theory of the mental is relevant to our understanding of quantum mechanics, or that there is no real multiplicity in nature, just the illusion of multiplicity in our minds" (Wallace 2012b, 3). Now it is crystal-clear that Wallace (2012a, c) is not a subjectivist in the sense specified within Box 2.1 (this is not surprising, given his 'membership' in the Oxford realist' school, but just reassuring with respect to this classification). It is also clear that Wallace's theory cannot be *meant* to be in support of my above claim in favor of a non-existence of an objective time. Wallace' view is in fact similar to Zeh's (1999). So, what is *my* argument here? My argument is that Wallace's theorizing can be *turned* into a supportive argument—without changing anything in the formal development—if one *subjectively reinterprets* Wallace's theory, which is possible, given *my* theorizing in Box 2.1, and if one states that all decoherent histories are *subjective histories*. That all those histories only evolve within the mind. Although Wallace, as a realist, might not be happy with my reinterpretation of his theory, he might not be able to avoid it if my argument in Box 2.1 is applied. Thus, finally, the argument developed here turns out to be a more detailed version of the previous argument.

Let me summarize: The arguments provided in (a), (b) and (c) suggest that the time arrow might only be subjective. Since in the multiverse, different realities and different times would then objectively coexist, nothing is in the way, physically, for phenomena such as physiological anticipation occurring.

Box 3.1: The second law of thermodynamics and Schrödinger's coffee[21]
The second law of thermodynamics indicates that the entropy of a system should not decrease over time. Under certain plausible conditions (most conditions outside living beings if seen as 'separate entities') the entropy increases, and this process is often argued as not being reversible. Reversing it (if at all practically possible) would require such an amount of energy that, again, an overall increase of the entropy of the composite system occurs (like the effect of an air conditioner that is cooling a house and decreasing its entropy at the cost of a greater entropy increase for the environment).[22]

[20]More on 'many-worlds' versus 'many-minds' interpretations of quantum mechanics is to be found in Chap. 4.

[21]I am thankful to Tanja Schade-Strohm who suggested the Schrödinger's coffee idea to me in a discussion.

[22]This example was suggested to me by an anonymous referee of the book.

Hence, this law is often taken as a strong indicator for the existence of an actual time arrow (e.g., Zeh 1999) and would therefore form a strong argument against ideas such as parallel times or non-existent time. Thus, the second law of thermodynamics potentially is *the* big obstacle for an explanation of phenomena such as physiological anticipation. The second law of thermodynamics is *probabilistic* in nature (for a detailed explanation and instructive examples see Boltzmann 1895, 414–415). However, even if a system already starts at a point of high entropy (but not maximal entropy), the probability that entropy increases is always larger than the probability that it decreases. But such a difference in probability is more pronounced when the system starts at low entropy. Thus entropy increase is also driven by the starting conditions. Since we are living in a world with a high probability of entropy increase and spontaneous entropy decreases are hardly ever observed, and this seems to have been the case on our planet for millions of years and seems to continue to be the case, we are apparently living in a universe that started at very low, actually unlikely low entropy (called the problem of low entropy). (Several solutions for this problem are discussed, but I would not classify the issue as settled; see, however, the interesting approach by Patel and Lineweaver 2015.) This clearly is a potential problem for those who desire using the second law of thermodynamics for the establishment of a time arrow; it is, however, not a problem within an adynamical explanation of the universe (see, e.g., Silberstein et al. 2018).[23] But since I am neither advocating the flow of time together with an asymmetric time arrow nor an adynamical blockworld, the question is how to align the second law of thermodynamics with the Wheeler-DeWitt equation (DeWitt 1967) and the absence of time, the 'tempus nihil est' perspective or 'times as special cases of other universes.' In case of the coexistence of parallel times, different states of physical entities with respect to their entropy or 'history' would coexist. Think of Erwin Schrödinger having a cup of coffee close to his university that he prefers to enjoy with milk. He pours some milk into the coffee and then mixes the milk and the coffee with a spoon. If he moves the spoon slowly, even makes some little breaks, there will be many (shortly visible, even) versions of his coffee (before he drinks it), with milk and coffee mixed in a more or less perfect way. The different versions of the coffee coexist in a similar way as there are different versions of Schrödinger's cat. According to the multiverse perspective with times as special cases of other universes, each of these coffee cups resides in a different reality.[24] If consciousness were able to *somehow move* between different 'coffee-cup universes' and thus across different realities, consciousness should also be able to move *upstream*,

---

[23] I am grateful to an anonymous referee of the book who suggested this argument to me.

[24] Note that these are 'mental realities,' or 'perceptual' if you will, since I am opting for a version of the many-minds rather than many-worlds perspective in this book; see Chap. 4 for those different versions of the multiverse interpretation.

opposite to the normally perceived direction of the flow of time. Schrödinger could *in principle* opt for perceiving coffee cups with a *smaller* entropy than that of a cup where milk and coffee are perfectly mixed somewhat 'later' than the occurrence of the latter, assuming a 'regular' flow of time. Or in other words, consciousness would then be able to move in the direction of lower entropy.[25,26] Of course, I am not advocating consciousness to actually be able to perform 'tricks' such as this one. It is a *gedankenexperiment* aiming at setting the stage for physiological anticipation. The exact mechanism is not so important here. What counts is the principal possibility of such an action in the multiverse. For physiological anticipation to occur, some unconscious perception with an impact on the body would suffice.

But if the existence of parallel times can actually be justified (and defended against objections from decoherent histories or the second law of thermodynamics), isn't it at least possible to *sketch* some framework as to how *subjective experience of the flow of time* emerges? A graphic way of looking at *both* anticipatory reactions as well as our regular perception of a flow of time might be sort of 'lateral movements' of our consciousness between universes or realities or just 'locations' (such as in our example with Schrödinger's coffee). Perceiving some reality as 'later' than another might thus be seen as a *perceptual convention* more than anything else. One might even argue that such a convention is routed in *culture*. Many ancient cultures are known to have had a cyclic perception of time, such as the Incans or Mayas. The idea that time (or space) is nothing empirical but rather constructed can also be found with Kant[27]:

> Time is not an empirical concept that is somehow drawn from experience. For simultaneity or succession would not themselves come into perception if the representation of time did not ground them a priori. Only under its presuppositions can one represent that several things exist at one and the same time (simultaneously) or in different times (successively). (Kant 1996 [1781], A30/B46)

The view on time (potentially substantiated in future psychological or neuroscience research) presented here can be summarized as follows. We perceive time by taking 'snapshots' of different realities. Some 'perceptual convention,' cultural background, or a priori category in the sense of Kant tends to organize this flow of pictures in a way that it looks like a unidirectional flow of time (for inspiring,

[25]Note that no 'bodies' have to be moved, here, because the respective versions of the individual (and his respective coffees) are already there.

[26]Mensky (2010) has speculated along similar lines in different parts of his monograph. He uses this idea to understand the survival of living beings.

[27]For an overview of different classical concepts of time, see Grosholz (2011).

however, indirect empirical evidence for this see Gruber and Block 2012). Needless to say, that we do not always have to follow this convention. Since other times (including the future) are existing in parallel we might sometimes be able to anticipate future events.

'Meaning' or 'Purpose' of Consciousness

Most current theories of consciousness might be characterized as belonging to reductionist approaches where the work of consciousness is equated with some *specific cognitive operations* (a very pronounced example, for the case of philosophy, is Dennett 1991). In many cases, consciousness might not even be looked at when dealing with cognition or only as an epiphenomenon; a currently quite dominant paradigm in psychology, the *cognitive neurosciences*, actually an interdisciplinary effort between neuroscience and psychology, looks at representations of cognitive activities (e.g., problem solving, decision making) within brain activities (see, e.g., Andersen and Kosslyn 1992; Gazzaniga et al. 2013). Consequently, that research field might be described as a "field that is built on the assumption that 'the mind is what the brain does'" (marketing text for "Frontiers in Cognitive Neurosciences," Andersen and Kosslyn 1992).

Such type of theorizing (as well as the underlying empirical studies) would be categorized as belonging to the *easy problems* regarding consciousness, if dealing with consciousness at all and not just disregarding it, by David Chalmers (1995, 1996); whereas the *hard problem* of consciousness could be described by questions such as "why are some organisms *subjects* of experience?" or "why do *qualia* exist?" With qualia, contemporary philosophy refers to the subjective, the inner side of sensual experiences.[28] Directly opposed to Chalmer's hard problem argument is the already referenced approach by Dennett (1991) who simply denies that qualia exist, a perspective that is not very convincing to me and that might also be—I think—failed by introspection by others. Just touch your cup of tea or coffee and drink something from it, then you might already know what qualia is, first subjectively experiencing how it feels to touch your warm cup, then experiencing how it feels to have the liquid in your mouth.

Most people intuitively believe in the existence of free will (Nichols 2011) (see also Chap. 6). This applies to a variety of different cultures. As a more recent 'wound of mankind' (see Chap. 1), however, neuroscience appears to prove that this is an *illusion*; at least if free will is seen as being more than a last-second vetoing power by consciousness in motor control (see below). Perhaps feeling

---

[28]The term has first been defined in its modern usage by Lewis (1956 [1929]). See also Schade 2015, 334–335.

'forced' to understand that free will is not existing, objectively, perhaps for other reasons, many philosophers have chosen a *compatibilist* perspective (see, e.g., Dennett 2003b) holding people *responsible* for their actions under certain conditions despite a non-existence of *alternative possibilities* within their theory.

But then, what is the 'meaning,' the 'purpose' of consciousness in the sense of the hard problem, in the sense of qualia, why then should we possess it? As can already be conjectured from the usage of the words 'meaning' and 'purpose,' I am here going to employ a *teleological argument*.[29] "Questions about teleology have, broadly, to do with whether a thing has a purpose or is acting for the sake of purpose, and if so, what that purpose is" (Woodfield 2010 [1976], 1). Teleological or so-called design arguments have, e.g., been crafted in favor of the existence of God (e.g., Aristotle 1999 [350 B.C.], 5–6; Plato 2000 [360 B.C.], Timaeus 28a–34b; Aquinas 2006 [1265–1273], 19) or to disapprove philosophical positions such as the solipsism (Kant 1996 [1781], B 39 et passim). Qualia might also be a phenomenon where a teleological reasoning makes sense.

Phrasing the question a bit differently and applying the decision perspective pursued in this book, one might ask: Why should we subjectively experience anything if there is no effect of this basic feature of consciousness on our *decisions* whatsoever? Note that asking this question is inspired by the following related convictions: Consciousness is *not* a *byproduct* of physiological (brain) activity, it is not supervenient on the physical (see Chap. 5). And qualia, i.e., our conscious experience of life, are something *qualitatively* different from physiological processes. Those related convictions are radical departures from other well-known approaches (e.g., Lewis, 1994).

It is clear that the decision perspective taken in this book together with the teleological perspective taken in this chapter strongly suggest that consciousness might have the 'sense' of 'producing' something like free will (Schade 2015). Especially since the alternative perspective on *subjective experience*, watching of and acting in (with fixed roles) a technically advanced 3-D movie, with no possibility to change anything we see, is a view with not much teleological appeal.

Reinterpreting Libet and Followers

A newly inspired debate about free will and responsibility originated in the well-known Libet-experiments (Libet et al. 1982, 1983; Libet 1985). In those experiments, a respondent would report the moment when he made some conscious decision to be executed with a move of his hand. The surprising result was that the readiness potential in the brain for the motor action of moving the hand (measured via an electroencephalogram, EEG) was running ahead of the reported conscious decision. How, then, assuming a linear, irreversible flow of time, could

[29]See also Schade 2015, 341, footnote 31.

consciousness determine any choices? Whereas there has been a critical but inconclusive discussion about how to interpret those findings, e.g., by John Eccles (1985),[30] most interpreted them as evidence for (a) the non-existence of free will and (b) individual's perception of possessing free will being an illusion. This interpretation has, however, not been accepted by everyone including Libet himself. Since the observed order of events in the experiments has always been: (1) readiness potential, (2) conscious decision, (3) action, Libet (1999) argued that consciousness might still be able to *veto* the motor action. Whereas this argument has been criticized by, e.g., Velmans (2003) and Kühn and Brass (2009), recent experimental evidence seems to be in favor of this ability (Schultze-Kraft et al. 2015). However, vetoing an already prepared motor action is only a very limited type of choice, a very small subset of decisions we are intuitively thinking of when free will is concerned.

Meanwhile, the reductionist position obtained an important ally: results achieved employing a larger modification of the Libet experimental paradigm, i.e., neuroscience studies allowing subjects to actually *choose between two alternatives* (i.e., pressing a left or a right key) (Soon et al. 2008). Here, consciousness has not only been demonstrated to run *several seconds* after specific activities in the brain. But, based on *specific brain areas* that were activated *before* the conscious decision was reported, the authors were able to *predict* respondents' choices. When a certain brain area would be activated, the individual would make a choice for, say, left, a few seconds *later*, reported as a choice made within consciousness, and after that the person would press the left key. The same would hold for the decision to press the right key. However, a *different brain area* would be activated ahead of time. Now it was crystal clear, the researchers argued, that consciousness *choosing* anything (beyond, perhaps, some vetoing) is an illusion. "So how realistic is our perception of *free* voluntary acts?" (Schade 2015, 341).

This set of results is indeed very compelling *if* people believe in a linear flow of time with no possibility of any 'time-backwards' effects. Applying this framework that is so natural to most researchers, consciousness is simply running after the fact, and then it must be a byproduct of the physical activities in the brain. But what happens if we question this premise, if we take into account the detailed discussion earlier in this chapter, if we treat different times as existing in parallel ('tempus nihil est'), as special cases of other universes? If parallel realities as well as parallel times might grant us (i.e., our consciousness) with the possibility of *laterally* moving between different times (because they coexist), it is only a small step to suppose that consciousness has a 'backwards-directed' influence on what is experienced to what extent (see Chap. 4) by the individual. This in turn would allow for a very different perspective on the reported findings: The fact that the experience of a conscious decision takes place *after* building the readiness potential for a motor action, or *after* observable activities in certain brain areas, would become less important for

[30]The question how consciousness might *influence* (material) brain activities is further analyzed by Beck and Eccles (1992).

the free will debate; notwithstanding the fact that the exact mechanism how all this is supposed to work still has to be delivered. In any case, free will would then in principle be possible.

Box 3.2: Is consciousness irrelevant for the quantum measurement process after all? A critical discussion of the approach by Yu and Nikolic (2011)

Yu and Nikolic use experimental evidence to argue against any meaningful role of consciousness for the outcome of quantum measurements: "(…) it has been suggested that the consciousness plays a vital role for the quantum mechanics as it is capable of, and even required for affecting physical events through the collapse of the wave function" (Yu and Nikolic 2011, 931). The authors argue from a firm materialist standpoint, directly opposed to Heisenberg's position, but also Wigner's, Mensky's and Lockwood's. As can be derived from the above cite, their background is singular-universe (collapse) quantum mechanics. Thus, they want to, specifically, demonstrate that the *observer's consciousness has no impact on the collapse of the wave function*. They, however, include authors such as Mensky in their criticism because he also argues that consciousness is at the core of understanding the measurement problem, albeit without acknowledging that Mensky develops his theory within a multiverse perspective. Their argument is as follows. There are several experiments at the double slit (see, e.g., Zeilinger 1999) demonstrating that constructing a mere *possibility* to collect 'which-way' information is sufficient to destroy the interference pattern: "The existence of interference patterns depends solely on whether the 'which-path' information is in principle obtainable (…). Whether such information is registered in consciousness of a human observer, one can conclude, is irrelevant. Consequently, this conclusion leaves no other option but to reject the collapse-by-consciousness hypothesis" (Yu and Nikolic 2011, 935). There are three major objections to this argument as well as to the general thrust of their paper; I take the leeway in applying a multiverse perspective since Yu and Nikolic integrate multiverse authors such as Mensky in their criticism. First, and with a detailed analysis being unfortunately beyond the scope of this book, one might ask as to how some of the reported experimental findings would have to be reinterpreted if a *consequent wave-mechanics interpretation* were used instead of a particle-wave duality interpretation (that the authors follow). If particles were just seen as narrow wave packets (see Chap. 2 of this book), I would expect that part of the findings on the (non-) disappearance of interference patterns could easily be explained, quantum mechanically. But if consciousness is not needed, anyway, for achieving the observed patterns, then demonstrating its uselessness for this case has no meaning for the materialism/idealism debate. Second, for a 'multiverser' the minimum necessity of having consciousness involved in the measurement process is, following Everett's argument, the conflict between a wave equation in superposition, indicating a 'multiple reality' and the singular measurement outcome within our consciousness; and this conflict can only be solved by giving consciousness some *role* in the measurement process, so that a general devaluation of the role of consciousness à la Yu and Nikolic (2011) simply makes no sense in the multiverse. Third, an experiment, no matter which one, is put together and finally evaluated within consciousness if one follows the top-down-decoherence approach (Box 2.1). Thus, the question when and whether someone finally 'looks' at a sub-part of the (totality of experimental) results might not be seen as being at the core of the issue. Even the 'not-looked-at' results must be seen as happening within the whole measurement framework, starting within and ending within consciousness.

Do the Pieces of the Puzzle Fit Together? And What Can Be Learned for the Further Development of the Multiverse Interpretation?

Let me summarize: The goal of this chapter was to put together an interdisciplinary, 'presumptive-evidence' proof for the multiverse. This proof required a detailed analysis of what time is. And the way this 'proof' was supposed to be crafted was to select pieces for one puzzle out of several disciplines—with each discipline contributing one piece of the puzzle, to be selected out of the many pieces that the respective discipline might have contributed. Revisiting Fig. 3.1, let me start with *predictive physiological anticipation* (the upper left piece of the puzzle). This was enabled via parallel times. If times are parallel, it might in principle be possible for our bodies to anticipate 'future' developments, even though the exact mechanism as to how this might be achieved is not quite clear at this point. Moreover, having anticipated the future and—perhaps—having changed the present via some reaction of the body, only the multiverse grants us with a theoretical framework that does not lead to theoretical inconsistencies. One reality contains the changes in the present because of the anticipation of the future, one does not (As will be seen in the next chapter as well as in Chap. 6, in each of those realities some version of the individual resides, but with a different degree of conscious emphasis on them).

Having accepted the notion of parallel times and the possibility of 'backwards' (actually 'lateral') influences by consciousness, it is also possible to reinterpret the *Libet evidence* and the evidence from choice experiments carried out in the neurosciences (*philosophical positions regarding free will*; upper right piece of the puzzle). Other than the literature that often uses this evidence to reject the existence of an actual free will (in the sense of choosing otherwise, and beyond last-second vetoing of unconsciously prepared choices), nothing like this can be conjectured, anymore. Consciousness might be able to somehow ‘influence’ choices ‘backwards;’ or more precisely, put different emphasis on different realities (see Chap. 4). This argument holds even if the exact mechanism as to how consciousness might be able to do that is not in any way clearer at this point than the mechanism as to how physiological anticipation might exactly work. However, if consciousness has this ‘job,’ it is at least clear that we are able to provide an *explanation/role of consciousness* (lower left piece of the puzzle). Chapter 6 will, however, develop a view on the genesis of free choices that is helpful to get a better feel for the action of consciousness.

No matter how explanations for physiological anticipation and free will are going to eventually look like, I would like to argue that *parallel times* (or better times as ‘special cases of other universes’) are required for them to exist. But then, having ‘opted’ for parallel times or, frankly, for the non-existence of time (DeWitt 1967), it is clear that we have also entered a multiverse perspective (*interpretations of quantum mechanics*; lower right piece of the puzzle). In the ‘standard interpretation’ of quantum mechanics, involving a collapse of the wave equation (the most accepted version of the Copenhagen interpretation), parallel times are impossible; an actual collapse of the wave equation would lead to a strong notion of irreversibility. And the relational blockworld exhibits other problems—as has been discussed earlier in this chapter. Thus, all pieces selected mutually support each other. This is what the chapter was supposed to show.

Chapter 4 that now follows will concern itself with the comparison of *different multiverse versions* and will propose that version that will be underlying the philosophical and decision-theoretic developments in this book: the *clustered-minds multiverse*. Is it possible to ‘extend’ the proof of the multiverse and formulate some specific requirements for an appropriate version of the multiverse interpretation, too? Is it possible to come up with a ‘system specification?’ Yes, it is. One requirement is simple and will already be expected by the reader for various reasons, including the topic of this book and its purpose outlined in Chap. 1. But coming back to the ‘interdisciplinary proof,’ *possessing an actual free will* was argued to give our conscious experience, the existence of qualia, a *sense*. Thus, the multiverse version to be proposed should be free-will friendly, and, as will be demonstrated, not all are. The other requirement has already been discussed in some detail. Consciousness must be able to ‘travel’ between different times. Thus, different times must exist in parallel. This also hints at certain ways to interpret (and not to interpret) the multiverse interpretation. Both aspects require consciousness to be fairly ‘flexible,’ not to be supervenient on the physical. Chapter 5 will develop such a concept of consciousness. It might best be described via a *dualistic idealism*.

Chapter 4
How Different Versions of the Multiverse Interpretation Have Different Consequences for Free Will and Ontology: Developing the Concept of a Clustered-Minds Multiverse

Requirements of an Appropriate Version of the Multiverse Interpretation

Given the analysis in the previous chapters, the purpose of the book, as well as the state of affairs in the literature, there appear to be *five* basic requirements an appropriate interpretation of the multiverse interpretation should meet (additional, less important requirements as well as some criticisms of specific approaches will be discussed on the way).[1] Two of the four basic requirements were already developed within Chap. 3 of this book and will not be explained here again: The multiverse interpretation should be *free-will friendly* (Chap. 6 will add details on different philosophical positions regarding the free will problem and the exact type of free will that the multiverse grants us with) and it should—as a special aspect of the free-will friendliness—allow for the *parallel existence of different times*. Three more requirements stem from the philosophical literature on the multiverse: A multiverse interpretation should somehow address the *problem of probability* given the deterministic feature of the Schrödinger equation (see, e.g., Wallace 2012b, c). And it should be able to deal with the *preferred basis* problem (see, e.g., Wallace 2012a; Galvan 2010). The fifth is the *avoidance of ontologically irritating features* such as infinitely split minds or the 'mindless-hulk' problem (see, e.g., Barrett 1999, 186–192; see also Squires 1988, 1991; Lockwood 1996).

Whereas the fifth requirement will be developed in conjunction with the discussion of the different multiverse interpretations, the problems of probability and preferred basis will directly be addressed in the following. For a better overview, the five basic requirements are listed, together with short definitions and the respective source(s), in Table 4.1. Some of those requirements are, as it will turn out, connected in non-trivial ways.

[1]A preliminary version of this comparison, using less criteria and looking at fewer approaches, is to be found in Schade (2015), 343–351.

© Springer Nature Switzerland AG 2018

C. D. Schade, *Free Will and Consciousness in the Multiverse*,
https://doi.org/10.1007/978-3-030-03583-9_4

Table 4.1 List of requirements underlying the analysis of different multiverse interpretations

Requirement	Short definition	Source
Free-will friendliness	Consciousness being able to put different emphasis on different realities; necessary element within interdisciplinary search of evidence for the multiverse	Chapter 3 of this book as well as this chapter (to be extended in Chap. 6)
Parallel existence of different times	Necessary element within interdisciplinary search of evidence for the multiverse; required to solve the Libet problem—and thus to allow for a connection of consciousness and free will	Chapter 3 of this book
Addressing the problem of probability	How to deal with the fact that measurement outcomes appear to follow the probabilistic Born (1926) rule whereas the Schrödinger wave equation is deterministic?	E.g., Wallace (2012b, c), Squires (1991), Mensky (2010) (to be addressed in this chapter)
Addressing the preferred basis problem	Out of all the possible realities that might exist in the multiverse many are not quasi-classical; how are the quasi-classical realities selected?	E.g., Wallace (2012a), Galvan (2010), Lockwood (1996) (to be addressed in this chapter)
Avoidance of ontologically irritating features	Infinitely split minds or living alone in a world of zombies are examples for irritating ontological features	Barrett (1999, 186–192), Squires (1988, 1991) (to be addressed in this chapter)

The discussion of different multiverse versions within the next section sometimes does, sometimes does not follow a rigid schedule working through the five requirements with each approach—after its introduction. Indeed, there are good reasons to organize the discussion differently at times. However, the list in Table 4.1 gives an orientation (or check list) for the reader to be considered repeatedly. Throughout the analysis to be carried out in the next section, it will become clear that none of the existing multiverse interpretations meets all five requirements. The original interpretation by Everett (1957, 1973) will not be dealt with as a theory in its own right for a simple reason: It is quite hard to 'interpret' it, and part of what makes this chapter necessary is that different physicists as well as philosophers debate on how this should be done.

To prepare the comparison of different multiverse versions in the next section, the problem of probability as well as the preferred basis problem were announced to be dealt with in the remainder of this section. I will start with the *problem of probability*. This eventually is the problem of specifying some reasonable position with respect to the Born (1926) rule, successfully used in practical applications of quantum mechanics for many decades. It provides *specific probabilities* for different measurement outcomes. The justification of the Born rule is an issue within each interpretation of quantum mechanics, despite its superb empirical support (see below). Within a multiverse perspective, the problem of probability primarily arises from the fact that the Schrödinger equation is deterministic together with the fact that all elements of a superposition are realized in some reality with certainty. Greaves (2004, 425) separates the problem into two parts: (a) the incoherence problem and (b) the quantitative problem.

With respect to (a), "Everettian quantum mechanics appears to be a straightforwardly deterministic theory (…). (…) we can be certain that each possible outcome of a quantum measurement is realized in some post-measurement branch. In that case, can it even *make sense* to talk of probability?" Note that this is not the position that has necessarily been taken by other authors writing on the multiverse. Wallace, aiming at a decision-theoretic foundation of the Born-rule in many of his publications, searches for a way around that problem by *redefining* probability somehow. Mensky (e.g., 2005, 2007a, 2010), also in his search for free will (see the discussion in Box 4.1 below) as well as Albert and Loewer (1988), argue that *fractions of infinite minds* might be sufficient to justify probability. If, say, a quantum measurement may lead to a certain result with $p = 0.5$ and to another result with $1 - p = 0.5$, then consciousness is supposed to 'send' one version of the individual to reality 1, and then another to reality 2, ad infinitum. But why should this solve the problem? Eventually, there is an infinite number of versions of the individual, and each of them experiences the respective reality with $p = 1$. Note that Mensky's as well as Albert and Loewer's (1988) account has also been criticized by Adlam (2014) for conceptual reasons: "(…) if this strategy is followed, it can no longer be claimed that the Everett approach does not require us to add anything to the basic quantum formalism." Adding *nothing* to the Schrödinger equation is certainly a desirable property of a multiverse theory, however impossible to achieve at this point, from my perspective, as will become clear in the discussion of different approaches, below. Nevertheless, additions might be compared in terms of severity of actual changes with respect to the basic calculus of the Schrödinger equation. And transforming individual probabilities for measurement outcomes into fractions of infinite minds is almost as substantive a change as the collapse postulate of the standard interpretation.

With respect to (b), this problem would even remain if problem (a) were solved: "(…) can Everett recover probabilities that numerically agree with those of the Born rule? (…) when a superposition is unequally weighted, the instrumentalist (…) [quantum mechanics] algorithm would assign *unequal probabilities* to the various outcomes. In contrast, it seems that all Everett can say is that each outcome occurs in exactly one branch (…), which, we might think, will yield equal probabilities if any at all" (Greaves 2004, 425). Interestingly, as already mentioned above, it is not even clear, also outside the multiverse view (see, e.g., Landsman 2008) what exactly justifies the Born rule theoretically. After decades of different proposals, a few scholars have recently pursued ways to derive the Born rule from *human-centered* principles, either decisions in the framework of Savage's (1954) *subjective utility theory* (Everettian view: e.g., Deutsch 1999; Wallace 2012b, c) or *generalized probability theory* to be applied by human decision makers (quantum Bayesianism: e.g., Fuchs 2010).[2] Deutsch (1999) and Wallace (2012b, c), e.g., consider the application of certain normative principles or axioms such as perfect rationality or utility maximization in the sense of decision and game theory for the case of agents optimizing their 'play' in an Everettian world with 'successors.' Although all the authors', including Fuchs' (2010), theoretical development is based on human decision making, I would not be surprised if none of them would finally agree with the statement that the resulting quantum probabilities or even the resulting Born rule that they derive are 'subjective.'

But aren't they just that—*subjective*, applying those authors' research strategies? And wouldn't that quickly lead to a new problem—a problem that the authors most certainly try to avoid? A 'subjective' establishment of the Born rule would undermine the entire Everett theory since the wave equation is almost exclusively tested stochastically, i.e., via an application of the—hopefully—*objective* Born rule. (Empirical evidence for the wave equation is empirical evidence for quantum mechanics, after all, independent of the interpretation chosen and dominantly stochastic.) But that is not the only problem of the human-centered approaches in the context of the multiverse interpretation. The main problem here arises from the fact that the respective calculations circumvent, mathematically, the incoherence problem (a) without offering a satisfactory conceptual solution for it. "The point is that in a situation with no uncertainty, it seems that much of (…) [a decision-theoretic] framework cannot apply" (Greaves 2004, 437). So, this partially brings us back to problem (a), and both problems (a) and (b) are now jointly discussed.

Three lines of responses have been tried to those two problems, according to Greaves (2004), the last being his own argument[3]: (a) subjective uncertainty, (b) reflection argument and (c) fission-based interpretation of decision theory. All those arguments do not suffice as a solution of the two problems, however. The subjective uncertainty argument (a) is rejected by Greaves himself, on the basis that (given the long and convincing discussion in Greaves 2004, 438–442) "an Everettian facing an imminent quantum measurement has no right to feel uncertain" (Greaves 2004, 442). The reflection argument (b) is based on the versions of the individual after measurement (but before the measurement has been read out), being uncertain of their self-location.[4] Unlike Greaves, I personally do not appreciate this argument because uncertainty should only matter if it is decision-relevant.

[2]According to a very critical statement by Zeh (2013, 98), "(…) quantum Bayesianism is the most recent form of a shut-up-and-calculate mentality." What he wants to say with this is that the wave equation and the Born rule are just seen in quantum Bayesianism as some predictive tools without any explanatory power.

[3]The intensive discussion in the literature cannot fully be captured, here.

[4]This is an unlikely case with macro-world measurements, anyway.

This appears not to be the case, here. Specifically, why would it matter where I am? I would need this information for future decisions *if I were uncertain about them*, but here the argument becomes odd—do I have to again recur to ex-post selves? Here one appears to be entering a situation characterized by an *infinite regress*. Since the self-locating uncertainty would only matter if I experienced real uncertainty about the next period, there will never be a point where the uncertainty becomes decision relevant.

The fission argument (c) then *abandons* the notion of probability altogether and replaces it by the following rule (Greaves 2004, 430): "The rational Everettian cares about her future successors in proportion to their relative amplitude squared measures." Sure enough, probabilities are not needed within this definition (however, the same mathematical calculus to generate them is used), the discussion about uncertainty in a deterministic universe can be halted. But is this ad hoc assumption any better than just maintaining that the Born rule itself is based on mod squared amplitudes[5] and restricting it to predictions *along one decoherent history*, i.e., to the vertical case, whereas horizontally, as a decision criterion *between* different realities, one admits that the Born rule has nothing to tell us?

Note that an idea, somewhat related to that of Graeves, will be discussed with respect to the horizontal case below, mod squared amplitudes as generating the *degree of consciousness* allocated to a certain reality. This idea will also, somewhat hesitantly, however, be turned down for usage in this book. My personal take on the Born rule in the Everett world and its usage within this book is depicted in Fig. 4.1. In this view, the Born rule becomes an *auxiliary equation*[6] to the Schrödinger equation for the vertical case (relative frequency of measurement outcomes within one history) and will not be applied to the horizontal case (different realities) in any form.

Specifically, Fig. 4.1 is not showing more than that: Vertically, times (as special cases of other universes) organize subsequent measurements within one reality (better: consciousness is organizing perception using 'times'), and results (relative frequencies) will always resemble the Born rule of quantum theory. Horizontally, the Born rule makes no sense since all realities appear with certainty. Or from the perspective of the individual, he will have a version of himself in each of those realities. What is the consequence of this? With respect to the Born rule, I would, given the knowledge of quantum theory we currently (do not) possess, *not* ask for further proofs or theoretical developments; the proofs provided within the rational decision theory framework (e.g., Wallace 2012c) have already been shown to lead to conceptual problems. Justifying the Born rule, inside or outside the multiverse interpretation, rather remains on the agenda for future research.

---

[5]The calculation of them is a mathematical operation, applied to the wave function, leading to classical probabilities for different measurement outcomes (Born 1926).

[6]Herewith, I apologize to those mathematicians that might potentially see this as a misuse of terms. I personally feel that the term 'auxiliary equation' captures exactly what I want to say, here.

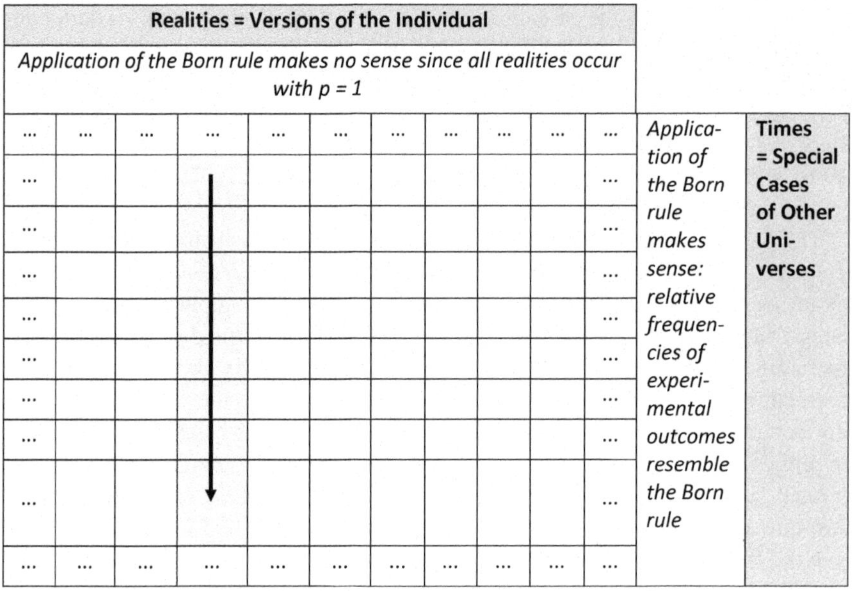

Realities = Versions of the Individual												
Application of the Born rule makes no sense since all realities occur with p = 1												
…	…	…	…	…	…	…	…	…	…	…	*Application of the Born rule makes sense: relative frequencies of experimental outcomes resemble the Born rule*	**Times = Special Cases of Other Universes**
…										…		
…										…		
…										…		
…										…		
…										…		
…										…		
…										…		
…										…		
…	…	…	…	…	…	…	…	…	…	…		

Fig. 4.1 Applicability of the Born (1926) rule, vertical and horizontal, in the Everett interpretation. *Note* The arrow within the grey-shaded area depicts the time horizon of measurements taken (simplifying, just the number of quantum experiments is relevant) within one decoherent history, experienced by one version of an individual

Box 4.1: Free will via an impact of consciousness on subjective probability? A critical discussion of the approach by Mensky

Mensky (e.g., 2005, 2007a, 2010) has developed a concept of subjective influence on quantum probabilities that he also considers the basis for free will. His idea will be critically discussed in this box: "(…) consciousness can increase the probability of finding its way into those classes of Everett's worlds that are preferable to it for some reason. This assumption may seem to be unacceptable when the probability of an alternative is identified with the fraction of Everett's worlds of the corresponding type (in which this alternative is observed). (…) the number that expresses 'the fraction of the worlds of a given class' should be universal, and must then coincide with the quantum-mechanical probability and may not be different for the consciousness of one observer or another" (Mensky 2005, 404). Mensky then offers a mathematical argument to overcome the problem described in the above reference by exploiting the fact that the number of Everett worlds, in his view, is infinite: "(…) the number that expresses 'the fraction of worlds of a given class' is meaningless for the infinite set of worlds (…). The reason is that an infinite set possesses a paradoxical property: it may be put in a one-to-one correspondence with its own subset. That is why (…) defining different probability distributions on this set is quite admissible and the assumption of the effect of consciousness on the probability distribution is not self-contradictory" (Mensky 2005, 404). There are a few problems with Mensky's theory of free will, however. The identification of part of the problem requires repeating, rephrasing and extending what has just been discussed with respect to the problem of probability: Each individual ends up splitting into all those infinite minds whose *fractions* might, however, *feel completely irrelevant* to each later version of the individual ending up in just *one* of the infinite realities; it might not be irrelevant from some evolutionary, anthropic perspective, however (see, e.g., Mensky 2010), but this is not the topic, here. Freedom, as experienced by a decision maker, cannot be identified within this concept. The entire issue becomes even more severe because of the fact that the number of minds—for Mensky's reasoning to work mathematically—has to be literally *infinite* which has already been identified above as a strong deviation from the basic quantum formalism and has been as well as will be classified as ontologically irritating (see, e.g., Table 4.1). Being only less outspoken about his exact concept of free will in later publications (e.g., Mensky 2011), Mensky is unfortunately not solving this problem. Note that other, quite serious issues would arise if Mensky *would* actually turn away from his original idea of infinite minds and would actually move into altering the probabilities of measurement outcomes—in the sense of one probability for one measurement outcome. In Mensky's definition of free will, the matter is somewhat opaque: "What is free will? (…) all alternative behavior scenarios are present as superposition components but the subject can compare them with each other and increase the observation probabilities for the alternatives that seem more attractive to her" (Mensky 2007a, 403). This sounds as if *one* subject would reason on the observation probabilities for different measurement results with just *one* version of the individual. However, such an approach would not ontologically be more convincing as an infinite number of minds. It would potentially lead to the 'mindless-hulk' problem where conscious entities could be surrounded by zombies if different interacting individuals would have different preferences for experiencing certain realities (for a more detailed explanation see the discussion of Squires' concept below). A different interpretation of the last reference by Mensky (2007a) would be that there are as many versions of the individuals as there are different measurement outcomes. But then, as was discussed in some detail above, it is all but clear whether it makes sense to have probabilities associated with measurement outcomes that all occur with certainty, with some version of the individual (see also Fig. 4.1). To sum up, Mensky's approach of altering subjective probabilities does not offer a solution to the free-will problem, no matter how his approach might be interpreted.

The *problem of preferred basis* is subtler than the problem of probability. However, as with the Born rule, it is not only a problem within the context of the multiverse interpretation, but more generally for measurement in quantum mechanics as Lockwood (1996) points out:

> *Why* [do] things appear [in a certain way to us] (…) remains to be explained. This is known as the *preferred basis* problem. I must emphasize, however, that it is no more of a problem for the Everett approach than it is for the collapse theory. (Lockwood 1996, 167)

More precisely, the preferred basis problem can be defined as (a) the *fact* that the Schrödinger equation does not only *allow* for *quasi-classical* realities but in fact for all kinds of realities, including non-classical, quite unusual ones, too; and (b) the resulting question as to why we are (normally) *perceiving* quasi-classical and not those other possible worlds (see, e.g., Lockwood 1996, 166–167; Bacciagaluppi 2001; Wallace 2012a). For a long time, the problem has pragmatically been avoided by defining "a phenomenological 'observable' that is used to characterize a measurement" in the laboratory (Zeh 2013, 101, explaining von Neumann's original approach) leading to a so-called "pointer basis" (Zeh 2013, 101). The same author as well as Wallace (2012a) propose a pragmatic solution of the problem via decoherence, that also applies to measurements in the macro sphere. They argue that because of the high degree of entanglement in the macro world, non-classical realities are 'filtered out.'

This sounds like a big relief and a solution to the problem. It is not, however. Wallace, suggesting a *physically self-organizing solution* to the preferred basis problem, notes himself that there are "far too many bases picked out by decoherence (…)" (Wallace 2012a, 63). Specifically, "(…) there are far too many system-environment splits which give rise to an approximately decoherent basis for the system; in the language of decoherent histories, there are far too many choices of history that lead to consistent classical probabilities" (ibid.). Referencing Dowker and Kent's (1996) strong criticism of this solution, he goes on: "Worse, there are good reasons (…) to think that many, many of these histories are wildly non-classical" (Wallace 2012a, 63).

While Wallace then turns to *computer simulations* of decoherence processes he perceives as successful (see Wallace 2012, 63),[7] others are generally more skeptical that the preferred basis problem is solved (within physics). An example is Jess Riedel's (Perimeter Institute for Theoretical Physics and former Ph.D. student of Wojciech Zurek) somewhat recent contribution to a discussion in physics stack exchange where he (Riedel 2013), repeating and summarizing Dowker and Kent's (1996) criticism that "*either* we need some other principle to selecting quasi-classical variables, *or* we need some way to define what an observer is without appealing to such variables" considers the preferred basis problem as *not solved.* Galvan (2010) theoretically analyzes the decoherence solution for the

[7]He argues that this solution is good enough for all practical purposes (=FAPP). However, such a FAPP solution may not suffice for addressing a severe *conceptual* problem like this. After all, results of simulations are generally dependent on the specific setup and the parameters chosen as well as the starting conditions selected.

case of the multiverse interpretation in detail and also reaches a negative conclusion[8]:

> The definition of pointer states is based on the decomposition of the global system into system + environment, and it is not at all clear how to perform such a decomposition *when the global system is the universe*; this is considered to be a severe conceptual difficulty of decoherence theory (…). The conclusion is therefore that decoherence does not even solve the preferred-decomposition problem in the context of the (…) [many-worlds interpretation].[9] (Galvan 2010, 8; italics by the author of the current monograph)

All this is consistent with Lockwood's (1996) earlier position, who, especially for the case of the multiverse, proposes a *radically subjective view* on the preferred basis problem:

> (…) in the context of the Everett approach, I can see no good reason for supposing that the apparent macroscopic definiteness of the world is anything other than an artefact of our own subjective point of view. (Lockwood 1996, 170)

Lockwood (1996) sees many-minds theories as a way out of this problem, a position that will be discussed—and adopted, in a novel version, for the remainder of this book—later in this chapter. His continues with a deeper analysis, proposing his version of a many-minds approach:

> (…) in quantum mechanics, there is a fundamental democracy of vector bases in Hilbert space. In short, it has no truck with the idea that the laws of physics prescribe an *objectively* preferred basis. For a many minds theorist, the *appearance* of there being a preferred basis, like the *appearance* of state vector reduction, is to be regarded as an illusion. And both illusions can be explained by appealing to a theory about the way in which *conscious mentality* relates to the physical world as unitary quantum mechanics predicts it. (170)

The question as to how such a theory about "the way in which *conscious mentality* relates to the physical world" (ibid.) is supposed to exactly look like is one of the core questions to be analyzed within the remainder of this chapter as well as in Chap. 5.

Comparing Different Existing Multiverse Versions

Multiplicity of Physical Realities: The Realist Version of the Multiverse

It is debatable whether any of the recent multiverse interpretations is *fully* realist, and whether such a realist interpretation is the one that Everett (1957) originally had

[8]He also analyzes this solution within the framework of the Copenhagen interpretation (in the form of the reduction postulate) and arrives at a negative result, too.

[9]Galvan (2010) then proposes permanent spatial decomposition as a better solution for the preferred-basis problem. However, permanent spatial decomposition is "the (hypothesized) property of the wave function of the universe of decomposing continuously into permanently non-overlapping parts" (Galvan 2010, 12) which appears to be a niche theory that needs to be further discussed and developed elsewhere.

in mind.[10] Many would argue that the Everett theory rather leads to some type of a many-minds interpretation (Lockwood 1996; see below).[11,12] Leaving this open, here, the contributions that are typically seen as constituting a realist version of the multiverse are those of Saunders (e.g., 1993, 1995) and Wallace (2002, 2012a, b, c). They are often called *many-worlds* theories that "(…) identify the multiplicity of coexisting realities as a multiplicity in the *physical world*, such as the coexistence of different outcomes in a measurement, or a live cat with a dead cat"[13] (Bacciagaluppi 2001, 9). In many-worlds interpretations, the world is seen as *actually branching* into different worlds, observers as having a unique past but several futures, and consciousness as splitting together with the physical realities it is attached to (i.e., all of them), similar to cells in the case of reproduction by cell division (Bacciagaluppi 2001, 11). The preferred basis problem is addressed within the idea of *decoherent histories*. Specifically, decoherence is thought to automatically 'pick' quasi-classical bases for the different, splitting worlds (Saunders 1995; Wallace 2012a), a position that has already been criticized above.

Applying the check list presented in Table 4.1, this approach is *not* free-will friendly because it entails an entirely *passive* role of consciousness, a *fixed coupling of consciousness with the branching worlds*; observers are automatically splitting all the time and have no decision as to which world to experience to what extent. The realist approach tries to construct a *time arrow* out of decoherent histories. Note that the decoherent histories *create an objective path* in this interpretation, so that no room for parallel times remains (this has been discussed with respect to Wallace's approach in Chap. 3, but finally this conclusion was reached). (Note also, as has been stated in Chap. 3, already, that decoherent histories might be seen as *subjective* with entirely different consequences; see the below discussion of other multiverse interpretations.) The problem of probability is addressed via a decision-theoretic approach, trying to avoid the assumption of anything being

---

[10]In the words of one of the most vivid proponents of a realist interpretation, David Wallace: "(…) I neither know nor care whether I am describing the historical Everett's own view" (Wallace 2012c, 2).

[11]Perhaps the evaluation of the two different claims is related to the question whether Everett's original paper (or even his full dissertation [Everett 1973]) or DeWitt's interpretation of Everett's thoughts are taken into account. According to Lockwood, "(…) if one were to judge merely by the evidence of his published writings, one might be tempted to classify Everett himself as a many-minds, rather than a many-worlds theorist. For he never speaks of dividing or differentiating worlds or universes, but only of the 'branching' and 'splitting' of 'observer states'" (Lockwood 1996, 172).

[12]Zeh (2013) would also agree with classifying Everett as a many-minds theorist, and, according to my reading of his publications, this is also the interpretation he follows himself (see the section on the 'many-minds' interpretation).

[13]As has already been mentioned in Chap. 3's section "But are times really parallel?", Wallace avoids self-classifying as a many-worlds or many-minds theorist (Wallace 2012c, 3). As one of the major proponents of a realist version of the multiverse, one could have expected him to self-classify as a many-worlds theorist. And his (somewhat implicit) classification of Everett's theory as both 'many worlds' and 'many minds' (Wallace 2012c, 3) contradicts the view by Zeh (2013) (see the previous footnote).

needed 'in addition' to the wave equation (e.g., Deutsch 1999; Wallace 2012b, c), but this approach has been critically discussed, above. The solution offered for the preferred basis problem is, among others, critically discussed by Wallace (2012b) himself and his proposed methodology of simulations might not be viewed as readily convincing, at least for conceptual questions (again, see the last section). After all those negative evaluations, the only clearly positive evaluation can be made with respect to the avoidance of ontologically irritating situations. First, since consciousness of *each* observer branches in a fixed coupling with *all* possible developments of the world, all other observers that he sees in any of those infinite realities should be conscious, too. So, there is no room for 'mindless-hulk' situations (Barrett 1999). Moreover, the number of realities is *not* necessarily infinite to start with but evolves in an evolutionary process (it might become pretty large, though).

Hence, the evaluation of the realist interpretation of the multiverse appears not to be favorable. Four out of five criteria had to be evaluated negatively. One of those concerned the room for free will. Only the avoidance of ontologically strange situations was granted. An additional, specific criticism has been brought up with respect to the realist interpretation of the multiverse, and I am mentioning it here despite the fact that its evaluation is beyond the scope of this book: According to Albert and Loewer (1988), an ever-branching physical universe might violate the *law of conservation of mass*. Perhaps this is one of the reasons why Wallace (2012c, 3) avoids being classified as a many-worlds theorist (without opting for being classified as a many-minds theorist).

Multiplicity in the Mental Realm: An Infinite-*Minds Version of the Multiverse*

Albert and Loewer (1988) propose a 'many-minds view,' that is, in fact, an *infinite* minds view. "In many-minds theory, the multiplicity of coexisting realities is not at the level of the physical realm but of the mental realm" (Bacciagaluppi 2001, 5). Or in other words, not the physical realities are assumed to be branching all the time, now it is the minds that do so. Similar to Mensky (2005, 2007a, 2010, see Box 4.1), Albert and Loewer (1988) propose an infinite number of minds whose *proportions* of perceiving one or the other outcome of a measurement are assumed to resemble the *probabilities* of the "experimentally verified probability rule of quantum theory" (i.e., the Born (1926) rule; Squires 1991, 283, in an article comparing his and Albert and Loewer's (1988) view). So, if two outcomes of a measurement are, say, equally probable, half of the minds will see one of the two outcomes, and the other half will see the alternative outcome (see the more detailed discussion in the last section). The authors admit that "this talk of infinitely many minds sounds *crazy*" (Albert and Loewer 1988, 207). Squires (1991) adds that he is not sure "(…) that the idea of an infinite number of existing minds (…) makes ontological sense" (285). Lockwood's (1996) perspective is closely related to that of Albert and Loewer (1988), albeit he states to have arrived at those conclusions quite differently: "[The] (…) feature of positing a continuous infinity of differentiating minds is one that my version of the many minds view shares with that presented by Albert and Loewer (1988). Albert and Loewer, however, arrive at these minds by an entirely different route from my own, and assign them a different metaphysical status from that which they have in my version of the theory" (1996, 173). Given a similar outcome, however, I decided not to deal separately with Lockwood's and Albert and Loewer's approach and I am going to apply my check list jointly to the two approaches; with one exception: Lockwood offers an important, very different treatment of the preferred basis problem (see also the Lockwood quote at the end of the first section of this chapter).[14]

The infinite-minds interpretation is as free-will unfriendly as the many-worlds interpretation according to the realist view (an important departure from this, using the same basis, is proposed by Mensky (e.g., 2005, 2007a, 2010); however, Mensky's approach does not convincingly solve the free-will problem either; see Box 4.1). Again, there is no choice involved as to what reality to perceive to what extent. The existence of parallel times is not explicitly addressed in any of the two approaches. The validity of the Born (1926) rule is simply *assumed*.[15] The preferred basis problem is not at all addressed in Albert and Loewer (1988), but Lockwood (1996) proposes a subjective solution to it where consciousness decides on which type of reality (i.e., quasi-classical) to perceive.

There is no room for 'mindless-hulk' situations in the infinite-minds interpretation(s), since each observer's consciousness is supposed to reside in all realities. However, the oddity here, the strange ontological status, simply arises from the infinity of minds; together with the idea of 'fractions of minds' allocated to different types of reality this also marks a major departure from the Schrödinger equation as has been pointed out and critically evaluated, already, in the introductory section of this chapter as well as in Box 4.1 (when discussing Mensky's approach to free will); see also the discussion of this aspect within Mensky's EEC, following below. Therefore, the evaluation of the infinite-minds views turns out not to be more favorable than that of the many-worlds interpretation (the realist view), albeit for different reasons. Only Lockwood's treatment of the preferred basis problem might be seen as an important breakthrough. Altogether, the many-minds views might not be seen as the end of the theoretical development. Specifically, free will could still *not* be localized in any of those approaches.

---

[14]I am also not going deeper into Page's (1995) *many-perceptions* (or sensible quantum mechanics) theory. Although again ontologically different from many minds, the differences are too subtle to lead to a different evaluation with respect to the criteria than, e.g., Albert and Loewer's approach. Although assigning a subjective status to probabilities, keeping psycho-physical parallelism (Page 2011, 6) does not open any room for free will: "(…) there [is not] any free will in the incompatibilist sense, and consciousness may be viewed as an epiphenomenon" (Page 2011, 7).

[15]This, per se, is neither 'negative' nor 'positive,' given the fact that there is no satisfactory solution to the problem of probability, right now (see the discussion in the last section).

A Many-*Minds Version of the Multiverse: The View by H. Dieter Zeh*

Zeh (1970) not only developed the decoherence principle (albeit not under this name), but he is also one of the most vivid proponents of the multiverse interpretation of quantum mechanics as well as the many-minds version of it and has already been referenced several times in the previous chapters because of his great clarity and consistency in developing the consequences of the (almost[16]) sole prevalence of the universal wave equation. Talking about the role of the observer in the Everett interpretation, Zeh (2013) notes—and I consider this his *own preferred interpretation*:

> If unitary quantum dynamics applies universally, one obtains a superposition of different versions of all observers described by *separate wave packets in configuration space* – just as there are different observers (at different locations in space) in one classical world. Emphasis on this aspect has led to the name 'many minds' or 'multi-consciousness' interpretation, since the relative state with respect to the observer's mind describes the observer's 'frog perspective' of the quantum world (…). (101; italics by the author of this book)

Anyway, the observer is *passive* in this interpretation (Zeh 2013, 101), and, sure enough, "(…) the different observer states (…), whatever their precise definition, cannot dynamically feel the presence of the 'other worlds' that are described by the [relative] states (…) any more" (Zeh 2013, 102). There are no physically different worlds, but there is only one world, but different versions of observers have different views on it. Since the existence of such *subjective* Everett worlds within *one* physical reality is based on *dynamically decoherent realities*, their number is probably huge, but the number is not necessarily equal to infinity. Therefore, the term 'many minds' instead of 'infinite minds' might be justifiable, here, at least for pragmatic classification purposes.

Regarding the preferred basis problem, Zeh (2013, 101–102) argues: "The problem of how to define an objective pointer basis that is sufficient for all practical purposes was resolved by the theory of environmental decoherence (Zeh 1970; Joos et al. 2003); (…) the "normal" and usually unavoidable environment of a macroscopic system induces a preferred basis for the pointer variable or other quasi-classical property that is objectively characterized by its robustness against further decoherence." However, this argument is quite similar to the one put forward by Wallace (2012a), including the outcomes of his simulations, and can thus be criticized pretty much in the same way as specified above in connection with the discussion of the 'many-worlds' theory by Wallace.

Walking through the above-specified criteria, this interpretation still lacks the possibility to accommodate for free will (since consciousness is passively attached to the different realities), and, as already mentioned, the preferred basis problem is not solved in a way that I found convincing, already in connection with the very

[16]The validity of the Born (1926) rule still has to be assumed.

similar approach by Wallace (2012a, c). Ontologically, the theory does not pose any extreme positions such as 'mindless hulks' or infinite minds. The validity of the Born rule is just assumed, without any further qualification, within Zeh's framework.[17] Since Zeh (1999), within his own view, *constructs* an actual, almost objective, because of irreversibility (see also the discussions in Chaps. 3 and 5), 'arrow of time' from decoherence (see Chap. 3), it is unclear how he would finally look at the existence of parallel times within our practical, everyday domain of life.

Infinite Minds, but Consciousness is Not Passive: The EEC Version of the Multiverse

According to Mensky's multiverse interpretation, the extended Everett concept (EEC), consciousness is able to influence subjective probabilities in the way described and discussed in Box 4.1. In EEC, consciousness, instead of passively residing with all possibilities as in the realist interpretation as well as in the many-minds interpretation, gets an *active role*. This possibility is provided, according to Mensky, via a mapping of infinite possibilities, whose probabilities are provided by the Born rule, onto another infinite subset with *different* probabilities (according to the preferences of the subject). Note, however, that the implied splitting of infinite minds (transformed or not) is an assumption adding something more serious to the quantum formalism, something also not clearly confined to the domain of consciousness, but rather a change in the physical structure whose explicit modelling is still to be provided. As Zeh, Mensky assumes the Born (1926) rule to be given, establishing *objective* probability, as a 'starting point' for *subjective* influence to play out.

The fact that in Mensky's EEC, consciousness gets an active role and is assumed to execute free will, has to be applauded. However, over and above what has already been critically discussed in Box 4.1, there are a few more issues with Mensky's concept of free will:

(a) One issue is that Mensky only 'allows' the unconscious to have access to parallel realities (see, e.g., Mensky 2005, 2007a, b, 2010, Chaps. 1 and 2), a thought consistent with the fact that the evidence for individuals getting knowledge of the future, reported in Chap. 3, is regarding physiological (hence: mainly unconscious) measures (Mossbridge et al. 2012). But since Mensky's concept of free will requires *super-consciousness* (when awareness is turned off during sleep, meditation etc.) to have access to parallel realities (e.g., Mensky 2007b) and if consciousness then only sees *one* reality, how is it supposed to execute free will? One potential solution within Mensky's framework would

[17] Again, I neither classify this as a 'negative' or a 'positive' feature, given the status of knowledge with respect to the genesis of the Born rule.

perhaps be that the number of parallel realities that consciousness considers is, say, *smaller* than the number considered by the unconscious (or superconsciousness, for that matter), but sometimes larger than one.[18,19] Anyway, individuals might not often have access to future developments—no matter whether this is assumed to work via consciousness or the unconscious—in a narrow sense, as supposed by Mensky, but only in the 'regular' sense of more or less rational expectations in the macro world (see Chap. 8). In any case, conscious choices between alternatives could subjectively be experienced in the form of phantasies or 'case studies.'[20]

(b) Another issue is that Mensky's concept somehow equates perception with choice whereas in psychology, perception and choice are traditionally treated as separate processes (see, e.g., the textbooks by Hayes 1994; Lefton 1994). This problem will be addressed in Chap. 6. Sorting this out appears to be relevant also for the concept of the clustered-minds multiverse, i.e., the novel multiverse version that will be proposed below.

(c) Mensky does not offer a solution to the preferred basis problem in any of his publications. Within his formalism, he assumes the potential measurement outcomes to be given and clearly defined.[21]

Let me summarize the results for the EEC, consulting again Table 4.1. A free-will friendliness of his approach is asserted by Mensky, but his concept is problematic, given the above discussion and the considerations in Box 4.1. However, trying to give consciousness an active role and to establish free will is aiming in the right direction, even though the exact approach chosen by Mensky is unsatisfactory. The existence of parallel times is provided.[22] Regarding the problem of probability, Mensky assumes the validity of the Born (1926) rule in the

[18]'Sometimes' is an appropriate description since in many cases choices are surely made by the unconscious leaving nothing left to decide for consciousness (but this is certainly no case for free will).

[19]It will finally turn out that the execution of free will looks quite differently from what we expect (see below as well as Chaps. 6 and 8).

[20]More precise than the English term 'case studies' would be the German term 'Probehandeln' (better translated perhaps as: 'iterative mental testing of different possibilities to act') that had originally been used by Sigmund Freud.

[21]Note that Mensky's approach has been criticized in an earlier publication by the author of this book (Schade 2015) for another potential problem, the 'mindless-hulk' problem, where an individual might be confronted with other individuals that are *not conscious* (see also Barrett 1999). However, this criticism is only valid for some verbal statements made by Mensky, e.g., "In the simplest situation, a single alternative is selected" (Mensky 2011, 617). In his mathematical (and older) representations, Mensky (2005, 2010) is more careful in describing the way how consciousness is supposed to generate subjective probabilities. This approach, however, assumes infinite minds, splitting into fractions attached to certain realities, and it has been criticized in Box 4.1. It will also be critically evaluated below. The 'mindless-hulk' criticism does apply to one of the two suggestions made by Squires (1988) who is actually aware of the problem (see below).

[22]Mensky is very explicit about it and uses the notion of parallel times in his concept of post correction (2007b).

'objective quantum world,' as most other authors do (including, e.g., Zeh as well as myself (see below); myself, however, with the modification depicted in Fig. 4.1).

Two more disadvantages of Mensky's EEC lie in the two last criteria listed in Table 4.1, the preferred basis problem as well as the 'ontological oddity' problem. Both are simply disregarded in Mensky's reasoning. This is especially problematic with the latter (the preferred basis problem could perhaps be fixed, I suppose, in a way similar to what will be proposed within the novel multiverse interpretation, the clustered-minds multiverse, below): The number of realities is argued to be *infinite* in Mensky (2005, 2010), those publications containing explicit considerations on free will; and accordingly, there are fractions of those infinitely split minds allocated to different types of realities. The major modification this concept requires with respect to quantum formalism (still to be delivered) is the same as with the infinite-minds approach (see above). In Mensky (2011), he is not so explicit, anymore, about the number of worlds and the necessity of infinitely split minds within his theory, but this potentially leads to different problems, as has been pointed out in Box 4.1. Two additional problems have been mentioned, the role of the unconscious and the equation of perception and choice. All in all, Mensky's approach is very innovative, it is a very important departure from previous approaches of interpreting the multiverse interpretation of quantum mechanics, and it is inspiring for future research. Despite the problems analyzed above, my own approach, the clustered-minds multiverse to be presented below, is partially inspired by Mensky's work.

Free, but Alone—Or not Alone, but Unfree: Singular Perceived Reality and Universal Consciousness Interpretations by Euan Squires

Squires (1988) suggests two alternative interpretations of the multiverse. One is quite simple. Consciousness selects just *one* reality:

> First, in any situation where a random selection is normally made, there ought to be the possibility of making instead, a *specific* choice.[23] In this way we could perform "quantum" psychokinesis and cause a breakdown of the quantum probability law. In one sense, however, we would not affect the wavefunction. In the analogy of the T.V. set, if we knew the relation between the buttons and particular programmes, then we could *choose* which programme we observe. Why can we not do similar things with our observations?

Although appealing at first sight, ontologically, Squires was fully aware of the 'mindless-hulk' problem when suggesting the 'selection' of just one reality by an

[23]It is interesting to note that Albert and Loewer (1988) (besides introducing their many-minds approach described above), also suggest, as an alternative, an approach where consciousness *does randomly select one alternative*, applying the Born (1926) rule. Since this approach leads to no new insights regarding the topic of this book and also suffers from the 'mindless-hulk' problem, it was not described in more detail, here.

Table 4.2 Squires' (1988) 'selection of singular realities' suggestion and the 'zombie' problem (Schade 2015, 347, modified)

		Alternative realities	
		Reality 1: Mercedes	Reality 2: Tesla
Alternative individuals	Louise	Consciousness present	Consciousness absent
	Tim	Consciousness absent	Consciousness present

Note In Schade (2015), this criticism has been crafted against Mensky's EEC. This is, however, only correct for some verbal statements by Mensky; a different criticism applies to the more detailed analysis presented in Mensky (2005, 2010) (see Box 4.1)

individual. He then argues in a *teleological* way: "(…) how do we ensure that different observers see the same result? (…) I suppose I am here making the untestable(?) assumption that most people that I meet are conscious" (Squires 1988, 18).

Since the idea of individuals selecting one alternative is indeed appealing, a closer look at the 'mindless-hulk' problem, mentioned a few times so far, appears to be finally advisable. It is especially relevant with Squires' approach. To illustrate the 'mindless-hulk' problem, I am going to provide a simple choice example (see, for this example, Schade 2015, 347). A couple, Tim and Louise, jointly decides whether to buy a Mercedes or a Tesla. It is supposed to be the only family car. Louise prefers to buy a Mercedes; however, Tim wants to have a Tesla. Let me furthermore assume that each of them is *selecting* the preferred reality; from an alternative viewpoint, both are successful in *influencing the respective family choices* in their direction. Anyway, Tim's consciousness realizes buying and driving a Tesla, Louise's realizes buying and driving a Mercedes. This implies having generated two parallel realities. In one of these realities, Louise enjoys her life with Tim and the Mercedes, whereas in the other reality, Tim enjoys his life with Louise and their Tesla. The problem with this seeming perfect solution can be derived from Table 4.2 where Tim and Louise are listed in the rows, the realities comprising the two different cars in the columns.

There is no reality where consciousness of both individuals is present. "From now on, each of the two partners lives with a 'zombie,' since consciousness is turned away crosswise from the respective realities of the spouses. In this example, free will would be rather unlimited, but would have an extremely high price, too: to basically live alone" (Schade 2015, 347). This so-called 'mindless-hulk' problem has been explained earlier in quite some detail in Barrett (1999, 186–192) and it is also described by Lockwood (1996, 174–175).

I would like to commend Squires' judgment that being surrounded by 'zombies' does not only feel strange, but would intuitively not make much sense. Since an essential ingredient of the 'outside' are other conscious beings, the 'mindless-hulk' problem is related to the solipsism problem in philosophy where an individual cannot be sure of anything but his own existence and leads to a violation of any intuition of teleological 'rightness.'[24] (See Schade 2015, 347–348; this reference also pertains to the footnote)

Let me summarize the pros and cons of this approach, based on the criteria listed in Table 4.1. Squires' first proposal is a free-will friendly version of the multiverse. But it is quiet on the question of parallel times, the problem of probability as well as the preferred basis problem. Squires is aware of the ontologically strange, 'living-in-a-world-of-zombies' (or solipsism) consequences of his approach. Realizing the solipsism problem of his first proposal, Squires then makes a radically different second proposal; it is another interpretation of the workings of the multiverse:

> The only solution to this problem seems to be that "consciousness" has a unity, i.e., there is, in some sense, one consciousness which knows the result as soon as I (…) have made an observation. This universal consciousness must then guide the selection of any subsequent observer. (ibid.)

This solution, however, exhibits a new problem. It solves the 'mindless-hulk' problem by sacrificing any space for free will since the 'one consciousness' would have to kind of 'dictate' all individuals' measurements/choices (perhaps excluding the choice of the individual who *first* measured; but who would that be, in a world where all times are parallel?).[25] By requiring 'one consciousness' coordinating all individuals' measurements on one consistent picture of the world, Squires (1988, 1991) is "bringing back a *singular* reality 'through the backdoor'" (Schade 2015,

[24]Although (moderate) solipsism has been proposed by some including Schopenhauer, stating that 'THE world is my representation' (Schopenhauer 2010 [1818], 23), Kant, e.g., has strongly argued against such a position, actually using teleological arguments: "It still remains a scandal to philosophy and to the general human reason to be obliged to assume, as an article of mere belief, the existence of things external to ourselves (…) and not to be able to oppose a satisfactory proof to anyone who may call it in question" (Kant 1996 [1781], B 39). The following, somewhat humorous statement by Karl Popper shows how difficult this discussion in fact is: "I know that I have not created Bach's music or Mozart's (…) [,] I just do not have it in me" (Popper 1999 [1956], 83). Although this consideration nicely demonstrates that Popper simply cannot be *alone* in an absolute sense, it does not lend a clear support to other entities visible to him possessing consciousness. (See Fumerton 2006, for a great overview of different approaches to the problem of solipsism within philosophy.) Philosophically, the solipsism problem, to the best of my knowledge, has not finally been solved; perhaps this is impossible, in a narrow sense. But I have to admit that the solipsism problem (in the very basic form of individuals creating reality within their consciousness) cannot be avoided as a problem of theory, even, within any subjective solution to the measurement problem. The best solution, from my point of view, lies in the concept of shared reality. Other individuals *appearing* in the focal individual's *films* should be *conscious*. And this is what has to be secured in any subjective interpretation of multiverse quantum mechanics. Shared reality, the solution proposed here, lies at the heart of the concept of the clustered-minds multiverse (see below). This partial solution will, however, turn out to be much more comforting than a situation where the individuals simply *must* be alone, fully, as is the case with the 'selection of one alternative.'

[25]It would also have to dictate identical perceptions/choices of the preferred basis by all individuals.

348). All other characteristics from Table 4.1, parallel times, problem of probability and preferred basis problem are still not addressed, as is the case with his first proposal.

Introducing the Clustered-Minds Multiverse[26]

In this section, I will now introduce the concept of a *clustered-minds multiverse*, the interpretation of the multiverse that is proposed by the author of this monograph and that is underlying the remaining chapters of this book. It will turn out to solve most of the problems specified at the beginning of the chapter as well as the additional ones discussed within the analysis of different multiverse versions.

At the core of the problem of finding an appropriate interpretation of the workings of the multiverse is the question as to "how consciousness is assumed to be *distributed* between alternative realities" (Schade 2015, 349). All approaches that have been presented so far implicitly or explicitly assumed that consciousness either resides with just *one* reality or with *all* possible realities and is then equally distributed among them. Some approaches even assumed more minds than alternative realities: an infinite number, allocated to different measurement outcomes, split according to fractions resembling the Born (1926) rule. Consciousness residing with only one reality is a problematic approach since it leads to the 'mindless-hulk' problem, as has been demonstrated or requires a universal consciousness as a universal 'dictator.' With respect to the assumption of infinite minds, let me note that it is, ontologically, not only unappealing to have such a huge number of minds, but (implicitly assumed by the respective authors) that they are all *equally 'important.'* The situation is alleviated with many instead of infinite minds, but the equal distribution of consciousness might still be considered implausible.

So why are only the extremes considered: one reality gets all conscious emphasis, or all realities get 'equal fractions' of consciousness? (Altering the fractions of infinite minds, the approach by Mensky (2005, 2007b, 2010), does not change anything with respect to the importance of one mind, for one version of the individual.) But what is the alternative? The alternative is having densely and sparsely 'populated' universes in terms of the amount of consciousness allocated to them.[27] This situation is resulting from consciousness of different individuals being more or less clustered around certain versions of reality. This in turn requires, in terms of a theory of consciousness, the degree of consciousness, the emphasis that an individual's consciousness is putting on different experienced realities, to vary. The concept of different degrees of consciousness allocated to different versions of the individual might be understood via an analogy to the animal kingdom.

[26]A preliminary version of the clustered-minds multiverse, then under the name of 'densely and sparsely populated universes,' has been presented in Schade (2015, 349–351).

[27]I am very grateful to Tanja Schade-Strohm for suggesting this solution to me in a discussion.

Typically, different animals are viewed as possessing different degrees of consciousness (e.g., Griffin and Speck 2004). The fact that those levels might then be seen as being connected to a different complexity of their brains—whereas all versions of one individual have the same brain complexity—may not be irritating in the context of multiverse quantum mechanics; the reason being that according to dualistic idealism (to be proposed within Chap. 5) consciousness is nothing that is *directly* coupled to the physical (i.e., does not supervene upon the physical).[28]

Somewhat related to Greaves (2004, 430), one might speculate whether or not the degree of consciousness allocated to different realities could somehow be related to the *mod squared amplitudes* (see above), an interpretation that is less tautological than that of Greaves (2004, 430), just bringing in what is needed in Deutsch's proof, individuals caring "about her future successors in proportion to their relative amplitude squared measures." According to my interpretation, depicted in Fig. 4.1, the Born rule (prescribing those mod squared amplitudes) makes only sense within the multiverse interpretation in terms of a prediction of relative frequencies of the occurrence of measurement outcomes within one time line (i.e., one reality) (vertical perspective). But shouldn't the mod squared amplitudes have *some* meaning also horizontally, and couldn't that meaning be the degrees of consciousness involved with different realities? I have to admit that I find this thought appealing. Those, then, 'starting values' for the degree of consciousness residing with certain realities would imply that free will would make *adjustments* to this basic 'endowment.' This would be a very different usage of the Born rule's mod squared amplitudes within the free will debate than that proposed by Mensky (2005, 2010), critically discussed in Box 4.1.

Nevertheless, I have decided not to pursue this idea further in this book (with the exception of one footnote in Chap. 7, briefly exploring the consequences of this idea for responsibility). The reasons for the current dismissal of the idea are straightforward:

(a) The idea of the degree of consciousness being related to the mod squared amplitudes is extremely speculative without intensive further research/theory development; it is less tautological than Greaves' idea, but as speculative. In fact, there are neither any theoretical nor any empirical justifications for it at this point.
(b) Especially important within future theory development would be the question as to how the *same* mathematical calculation (mod squared amplitudes) leads to a prediction of relative frequencies of measurement outcomes within the vertical dimension (see Fig. 4.1) but to different degrees of consciousness within the horizontal dimension.
(c) Moreover, all realities (depicted horizontally in Fig. 4.1) occur with a probability of one, and there is a 'fully equipped' version of an individual in each of

[28]Of course, different degrees of consciousness are also considered with different cognitive operations of humans, with fully unconscious operations marking one end of the spectrum, and perhaps highly conscious deliberations the other.

them. Having a fully equipped version of the individual in some reality seems to me enough to assume that free will is able to put a *discretionary weight*, including a large one, on that reality—even if the mod squared amplitude attached to this reality should be small. But even if the idea of mod squared amplitudes setting the starting point of consciousness allocation would be accepted, how much adjustment by free will is possible; what are the adjustment rules?

Thus, I do not want to reject the possibility of a relationship between mod squared amplitudes and degree of consciousness, but I think that exploring this possibility in further research is a long-term endeavor by a number of researchers; most certainly, a valuable one, even if the idea would perhaps have to be turned down, eventually. Within this book, however, this is too premature an idea to be further pursued. The good news is that implementing this idea would not have changed much in the basic reasoning presented in the following chapters, I suppose, since free will would still be applied in the form of adjustments to the degree of consciousness allocated to different realities; having this new 'restriction,' everything would just become more complex.[29,30]

How might the concept of densely and sparsely 'populated' universes—in terms of the amount of consciousness allocated to them—be visualized? Let me first disregard the problem of preferred basis and illustrate this concept by using the simplifying *allegory of a torch light,*[31,32] whose cone of light is very bright in the middle, but the light intensity fades with increasing distance from the center. Let me further assume that an individual's consciousness is distributed pretty much in the same way as this cone of light. There is always a core reality where the *center of consciousness* resides, and then there are other realities where less consciousness resides. The 'distance' from the center is a distance in terms of choices that differ from the choices of that version of the individual that is associated with the core reality. Since this assumes multiple 'game rounds,' i.e., a history of decisions, the concept of decoherent histories (e.g., Saunders 1993; Wallace 2012c) appears most appropriate, here, albeit in the subjective, top-down form that was suggested in Box 2.1. Let me look at a situation where that version of the individual, located in the middle of the cone of light (the one associated with the core reality, where the

[29]I am not saying that complexity is per se negative, especially if the theory were correct. But, again, this is hard to be decided at this point.

[30]The largest difficulty would be the formal restrictions—to be specified—for the vectorial decisions (see formula (1)) introduced in Chap. 8. If, say, allocating away consciousness from a certain, negative reality is almost impossible because of a large mod squared amplitude that pertains to this reality, this would be a severe limitation, in that case, to free will and responsibility (see also Chap. 7). Free will and responsibility would become slightly more situation-dependent than with the theory currently pursued in this book.

[31]This allegory has the highest intuitive appeal with an old-fashioned torch light, since LED and laser have a more concentrated cone of light.

[32]I am thankful to Adam Taylor, who suggested the usage of the torch light allegory to me after my presentation at IINN's Free Will Conference in Flint, MI (2015).

center of consciousness resides) decides to make a left turn at some traffic light using his car. In the multiverse, there will always be a version of this individual making the right turn. Now, the version making the right turn is slightly away from the center, with reduced consciousness. The more the choices in the 'choice history' of a certain version of an individual differ from the choices of the 'center version' of that individual, i.e., the larger the above-defined distance from it, the less bright will be the light of the torch shone on it, and consequently, the lower the amount of consciousness allocated to this version. "In other words, there is a smooth removing of consciousness from realities that are close to the 'center (…) [version],' a strong removing of consciousness, however, from those that are located 'many decisions away'" (Schade 2015, 350).

A similar allegory has been proposed within the philosophy of time: the 'moving spotlight.' "According to the moving spotlight theory of time, the *property of being present* moves from earlier times to later times, like a spotlight shone on spacetime by God" (Skow 2012, 223). The 'moving spotlight' theory assumes the parallel existence of different times within spacetime (see, for a similar notion, the block universe view that has been discussed in Chap. 3), a concept that differs from the one preferred by the author of the current book: 'tempus nihil est.' Then there are other differences: Not only times are parallel but also realities, at each point in time, separated by decisions or measurements.[33] Also, each individual is using a separate torch light, whereas there is only one moving spotlight, assumed to be 'universal' leading to an *absolute* past, presence, and future (Skow 2009, 2012). Finally, whereas the time-spotlight has sharp boundaries and shines on just *one* time (such as a laser pointer would), the torch light in my allegory shines on many realities; the intensity, however, diminishes with higher distance from the center.

The torch light offers a simplifying picture, helpful to get acquainted with the concept of distribution of consciousness between numerous versions of one individual. It is *too* simple, however, since a smooth and continuous reduction of consciousness, looking around from the place of the center individual, is implausible. *Some* 'outer versions' might actually be shining bright (with a lot of consciousness allocated to them) and the resulting light cone might be *blotched*. Also, it is implausible that only the 'center individual' has the power to make choices (and 'drags' all the others along), especially if some 'outer versions' are indeed highly conscious. It is much more plausible that each of the versions has some (perhaps very small) influence on where the light cone moves, the influence being the greater the larger the amount of consciousness residing with the respective version is.

I have now reached a point where it is important to consider the problem of preferred basis within this framework. The first part of the solution of the problem has already been suggested by Lockwood (1996) and discussed above. His solution to the preferred-basis problem is a *subjective* one. To put his solution into my own

[33]Decisions in the multiverse will turn out to differ quite a bit from our intuition (see Chaps. 6 and 8).

words: Consciousness *normally prefers* to see quasi-classical realities.[34] Lockwood's subjective perspective also solves the preferred basis problem within the clustered-minds multiverse if one assumes that there is a *smoothing effect of the torchlight* not only on different realities in the sense of alternative, pre-specified measurement outcomes but also on the *perceptual* (or perspective) differences that the subjective selection of preferred basis by different individuals might lead to. Clearly, in all subjective, minds-based interpretations of quantum mechanics two individuals never reside in *exactly* the same reality, because their respective consciousness introduces a *specific subjectivity* into their *perception* of the world. Thus, two individuals 'agreeing' on their 'mutual appearance' in the respective other's world, have to share a *similar perspective* to be able to communicate. This situation might perhaps also be solved via the fuzzy, smoothing effect of a non-punctual consciousness typical for the clustered-minds multiverse. This should apply also within much larger groups of individuals: *Perceptions should be sufficiently similar within a minds cluster.*[35]

There is a final, very important point that should be made with respect to the preferred-basis 'problem.' There is an aspect that actually turns that problem into a *preferred-basis solution*. The preferred-basis solution is a partial solution to the problem of too many realities. If individuals subjectively 'agree' on certain ways to view the world within consciousness clusters, it is implausible that they use an infinite number of such perceptual 'platforms,' i.e., form an infinite number of such clusters. Thus, this is the most plausible way to reduce the funny plethora of possible realities arising within other multiverse interpretations from the plethora of different possible preferred bases.

The clustered-minds multiverse builds up upon the following eight premises and has the following characteristics, respectively:

1. The multiverse is based upon the Schrödinger wave equation (or alternatively, the Wheeler-DeWitt equation) and allows for all kinds of *perceptual*, i.e. subjective, *possibilities*, in principle also odd, non-quasi-classical ones. In terms of physical reality, no implicit or explicit additions to the wave equation are needed. However, within repeated measurements *along one decoherent history*, frequencies of measurement outcomes are governed by the Born rule. This rule is thus seen as part of the quantum formalism in the sense of an auxiliary equation.

[34]This might not always be true. People with certain psychiatric conditions might be viewed as having opted for perceiving non-classical realities. The same applies with certain drugs (see Chaps. 6 and 13).

[35]I would like to leave open here the question as to what extent the phenomenon described within this paragraph is mainly organized within the wave function or mainly (or even solely) within the domain of consciousness. One might be reminded of Squires' 'universal consciousness' required to coordinate individuals' different perceptions of the world, albeit with far less 'dictator power' and much smaller reach. Not everyone is supposed to see the same reality, but individuals in a minds cluster are supposed to share their perception of the world.

2. There is only *one* physical reality (i.e., the reality characterized by the wave equation), but different experienced realities exist within consciousness. These are the (subjective) decoherent histories that describe our lives. According to Box 2.1, *decoherent histories start in consciousness* (top-down view). The clustered-minds multiverse is a descendant of many-minds rather than many-worlds theories. It is a *subjective interpretation of quantum mechanics.*
3. All times exist in parallel, or, equivalently, time may be seen as not existing at all: 'tempus nihil est.' No matter what phrasing is used, 'times' are special cases of other universes. The flow of time is as much an illusion within consciousness as is the uniqueness of experienced reality.
4. Free will is executed via an allocation of different degrees of consciousness to different realities (see Chap. 8, for a formal definition of so-called vectorial choices). Two extremes are excluded: Putting all emphasis on one reality, and putting no emphasis on a possible reality; each possible reality gets at least a marginal amount of consciousness allocated to it.
5. In order to maintain different minds clusters, somewhat balanced distributions (e.g., a lot of consciousness resides with, say, fifteen realities, all other realities receive only a small amount of consciousness) are normally preferred over more extreme ones (such as, one reality gets as much consciousness as possible, all other realities get only marginal amounts; an allocation like this would bring back sort of a singular reality and would potentially lead to numerous very-low-consciousness-situations, to be avoided by others; see point 8 of this list of premises).[36]
6. The preferred-basis problem is solved via consciousness mostly selecting realities that qualify as *quasi-classical*. Moreover, the preferred-basis problem is turned into a preferred-basis solution for the potential 'ontological oddity' of too many realities (arising from the plethora of different possible preferred bases). Consciousness clusters may decide in favor of experiencing several, but, most likely, not a ludicrously high number of possible realities.
7. *Very-low-consciousness situations* are actively avoided in the sense that individuals do not allocate a lot of conscious emphasis to situations where nobody else does.[37] This automatically avoids the emergence of 'mindless-hulks'

[36]I would like to argue that this type of a 'soft restriction' is not under the sole discretion of the decision maker, although the decision maker has many good reasons not to choose extreme allocations on its own (see Chap. 9). Instead, minds clusters might have an influence on the extremeness/moderateness of allocations. This factor has to be better understood in future research and will not often be referred to in this book. However, responsibility considerations require taking that factor into account (see Chap. 7).

[37]For the sake of simplicity, I will assume that *folie-à-deux* or *folie-à-trois* situations where versions of two or three individuals that possess a large degree of consciousness reside in a reality where all others hardly 'left' any consciousness are impossible, too, but this is debatable. It is clear, however, that not every consciousness cluster is necessarily large. The open question is: What should be the minimum number of highly conscious versions of individuals required for a consciousness cluster?

(or better: close-to-mindless-hulks) situations.[38] This all may sound like an odd assumption, at first glance, but the opposite, producing very sparsely 'populated' parts of the multiverse with just one entity being highly conscious (that version of an individual in turn dealing with almost robotic other entities) and the highly conscious version not realizing this somehow or even feeling 'comfortable' there, appears stranger to me. (This assumes that we have kind of an intuition for being 'alone' or better: 'almost alone,' and this is certainly a speculative thought, but this idea is not contradicting quantum mechanics; this ability pertains to the domain of consciousness.) This seems to require a coordination between different individuals' consciousness, but not in the radical sense supposed within Squires' (1988) universal consciousness interpretation. Individual freedom remains.

8. Since the emphasis that consciousness puts on different versions of the individual differs and since preferences of experiencing certain realities differ between individuals, free will normally leads to an imbalanced distribution of consciousness between different realities. The result is a coexistence of scarcely and densely populated universes containing clustered minds with more or less emphasis allocated to them by different individuals.

Further exploring the last two points, we may either find ourselves in rather densely populated universes, defined, say, as a cluster of 'similar realities' or simply a *minds cluster*, where a lot of consciousness from many individuals resides; the condition being that many individuals have made *decisions* to allocate a large amount of consciousness to those realities. Sometimes, however, we are going to find ourselves in rather sparsely and/or heterogeneously populated universes, where a few people, say, have made choices to allocate a lot of consciousness into this reality; but where others might have decided to be involved to a smaller degree. If such situations are too extreme, in terms of contrasts between degrees of consciousness, they are typically avoided. In any case, and part of the premises specified for the clustered-minds multiverse above, free will is not able to produce situations with realities where a conscious version of the individual is surrounded by almost robotic entities. Moreover, no realities are completely idle of consciousness. Even 'shadow realities' contain versions of individuals with a marginal amount of consciousness allocated to them.

Let me now apply Table 4.1 to this novel concept. First of all, I would like to argue that the clustered-minds multiverse is free-will friendly because individuals decide on how much emphasis they put on different versions/different realities (under some restrictions, however). The clustered-minds multiverse is based on the notion of parallel times or 'tempus nihil est;' the flow of time is seen as a purely subjective experience. The solution of the preferred-basis problem is at the core of the concept, as well as the avoidance of 'ontologically irritating' situations such as 'mindless-hulk' or 'infinitely-split-minds' cases. The Born rule is interpreted as an

[38]In narrow terms, mindless-hulks situations are avoided already by the fact that marginal consciousness has to be allocated to each reality (see above). Herewith, encountering close-to-robotic entities is also excluded.

auxiliary equation (thus part of the quantum formalism) for 'vertical,' along one 'history' measurements. The Born rule has no meaning for horizontal perspectives concerning alternative realities since they all occur with a probability of one. Nothing else has to be assumed in terms of additions to or modifications of the Schrödinger (or Wheeler-DeWitt) equation. Although many of the issues have to be explored more in several of the remaining chapters of the book, I would like to argue that a start is made, that the clustered-minds multiverse is an interesting variant of the multiverse interpretation of quantum mechanics, addressing a few important issues in a novel fashion without contradicting the quantum formalism.

PART II
ANSWERS SUGGESTED BY THE CLUSTERED-MINDS MULTIVERSE FOR SELECTED QUESTIONS IN PHILOSOPHY

Chapter 5
Dualistic Idealism: No Supervenience of Consciousness on the Physical

Mind and Body—Different Realms?

With respect to the mind-body problem, philosophy has seen a debate over many centuries, actually millennia (for an introduction into the current state of the debate see, e.g., Kuczynski 2015).[1] Since detailed reviews of the numerous positions that have been taken are beyond the scope of this chapter (and this book), I would like to simplify matters by starting with three stereotypical positions, two of them marking the two most extreme views that can be taken. One extreme, *Indian idealism* in its different variants (see, e.g., Dasgupta 1962), a type of *monism* (i.e., the world is made of only one 'substance'), can be summarized under the heading that only the mental exists, and that the rest is an illusion. A *dualist* would take a moderate position and claim that both the physical and the mental exist and that they are different realms. He, however, would have to address the question as to how exactly those two play together.

In mainstream contemporary philosophy, another extreme, a form of monism that is opposite to Indian idealism, is dominating the analysis of the mind-body problem: *physicalism*. Physicalism also dominates life-science disciplines such as psychology and biology. In a way, a physicalist either states that there is nothing but physical laws (or even just matter!), or he argues that all else is *directly* following from them. Dualist perspectives arguing in favor of a special realm of the mind or a special role of consciousness are typically dismissed because, from the perspective of a physicalist, mind or consciousness simply have no function; *consciousness is seen as an epiphenomenon.*

A group of philosophers has developed a *physicalist inter-pretation* of the quantum multiverse (the so-called 'Oxford interpretation;' see, e.g., the contributions to Saunders et al. 2012; Wallace 2012a, b, c) that has been presented

---

[1] I do not, however, follow some of the consequences Kuczynski (2015) derives from his analysis. A thorough discussion is beyond the scope of the book.

© Springer Nature Switzerland AG 2018

C. D. Schade, *Free Will and Consciousness in the Multiverse*,
https://doi.org/10.1007/978-3-030-03583-9_5

and discussed, already.[2] The respective authors call the Oxford interpretation a *realist* interpretation, and this label was also used in the last chapter, but it is also a physicalist interpretation. Physicalists mostly associate the multiverse with an *actual* branching structure (often based on decoherent histories) where new, physically distinct universes are *produced* with each branching. It has already been mentioned, however, that the most well-known proponent of the Oxford group, David Wallace, is somewhat unclear in his monograph (Wallace 2012c) with accepting or not this perspective on his branching theory; he also appears to avoid the consequences of not being associated with this perspective, however (see the previous chapters).

There are good reasons to be unsatisfied with the Oxford interpretation, despite all its merits, as could be shown in the last chapter. Continuing the discussion, here, instead of seeing the multiverse interpretation of quantum mechanics as a platform to better understand the workings of consciousness,[3] the fact that there is no collapse of the wave function, triggered by conscious observation, is surprisingly interpreted as an important step towards an *objective* description of the world. However, especially in the multiverse the question what exactly will be observed by whom, or how consciousness is *distributed* between different realities, should catch our attention as has been made clear, already, in the last chapter. Assuming an objective branching of the universe with consciousness passively coupled to all branches might not be judged as the most convincing route to take, ontologically.

Following my own interpretation, here, the clustered-minds multiverse, I should thus emphasize the oddity arising from the fact that it is indeed physics itself that poses the biggest challenge for a physicalist's worldview. The perspective offered in the clustered-minds multiverse or within any other *subjective* interpretation of quantum mechanics (that, as was derived in the previous chapters, is inevitable to address some of the issues such as the preferred-basis problem) is so different from the perspective that arose within classical physics a few centuries ago and so challenging, too, that our worldview has to change radically also with respect to the mind-body problem.

[2]Note that not all contributions to Saunders et al. (2012) are actually in favor of the Oxford interpretation, and some are even skeptical with respect to the multiverse interpretation of quantum mechanics.

[3]An early treatment of the mind-body problem from the perspective of the multiverse has been proposed by Lockwood (1991). However, his treatment of the problem and mine differ in various aspects allowing only for partial fertilization of my developments with his thoughts. E.g., he supposes an *identity* of mental states with brain states (e.g., 71–72) whereas I see the brain as just one stage in the entanglement chain not necessarily connected in any *direct* way with the mental states. On the other hand, there are also similarities between our approaches such as an orientation towards many-minds instead of many-worlds theory (see also Lockwood 1996) or using the picture of a light cone, albeit differently (see 73–75 of his book as well as Chap. 4 of mine). See also his interesting discussion of *realism* and beyond in light of quantum measurement (219–239).

Box 5.1: Is 'physics' the stationary state of the closed system (the universe) or the 'inner, subjective workings' of the open system (containing observers)?

Later in this chapter, I will address the question as to whether or not consciousness has an impact on the physical. This requires a definition of what should be defined as 'physical.' There are two possible ways of doing this, based on some of the issues discussed already in Chap. 2 (see also the literature referenced there). One is to take the perspective of an *outside observer* of the universe as a closed system (that is in a stationary state). The other is to take the perspective of *inside observers* and to label as physical the decoherence within the open system, that was stated to be caused by consciousness (see box 2.1). For me, 'subjective physics' is 'real physics,' at least the subjective perspective decides on what we consider as 'real' (preferred basis) and how much emphasis we put on different measurement outcomes in our experiments or just on different realities (see also Chap. 13). Therefore, even though it could be argued that consciousness has no 'objective' effect on the physical, looking from 'outside the universe,' it does have an influence on the world we experience.

This chapter is organized as follows. In the next section, I am going to revisit the evidence already collected within the last chapters (with a special focus on the issues considered in box 2.1) pro a top-down interpretation of the quantum multiverse, i.e., from consciousness to quantum, that I am advising, most naturally, also in the context of the mind-body problem. In the third section, a core concept in the contemporary philosophical treatment of the mind-body problem, the concept of *supervenience*, will be introduced. Supervenience is what physicalists want to prove and non-physicalists want to disprove with respect to the relationship between mind and body. The core concept underlying the contemporary dualist view on consciousness, the hard-problem analysis of consciousness, will also be dealt with. The fourth section will then build up upon the second and develop a multiverse view of the mind-body problem and show that *supervenience* of consciousness on the physical can be *rejected* in the clustered-minds multiverse. However, building up upon the definition of the 'physical,' just proposed in box 5.1, and again recurring to the top-down perspective from box 2.1 and other sections of the current chapter, I am going to show that consciousness *does* have an influence on the physical—and this without running into some of the problems that have been mentioned in the literature for that case. The last section will finally summarize the consequences of my analysis for the mind-body problem and introduce the concept of dualistic idealism. It will finally discuss an alternative perspective, weak psychophysical parallelism, in box 5.2 at the end of the chapter.

Bottom-up or Top-Down Creation of Reality in the Multiverse?

In the previous chapters, the question *where reality starts* has been encountered several times. In box 2.1, I have already opted in favor of a *top-down view*. Although partially repetitive, I will here summarize some of the arguments provided in box 2.1 and elsewhere for the context of questions relevant for this chapter.

Some authors try to explicitly build reality *bottom-up*, starting in the world of *decoherent histories*. This is a typical feature of the *realist interpretation* of the multiverse (e.g., Saunders 1993, 1995; Wallace 2002, 2012 a, b, c). The bottom-up approach leads to numerous unresolvable issues, however, that make it impossible, from my point of view, to pursue this approach successfully. Important are the *preferred basis problem*, with far too many possible bases when starting in the 'physical' domain (see Chap. 4). Also troublesome is the *conservation of mass* problem if a multiverse version with an actual splitting of the universe is opted for. (This is, however, not relevant within the many-minds approach by Zeh (2013), for sure, whereas Wallace, e.g., 2012c, as already stated several times above, e.g., in the related form of an actual versus a subjective branching structure, currently keeps the question of many minds versus many worlds 'open.') Finally, and differing from Zeh (1999),[4] I am not convinced that an *actual* arrow of time can be *constructed* from decoherence (as has been discussed, already, in Chap. 3).

The opposite strategy is to construct reality *top-down*, i.e., starting in the domain of *consciousness*. The basic idea that consciousness rather than 'physics' should be the starting point of quantum reality has first been suggested by Wigner (1983 [1961]) in the context of the Copenhagen interpretation of quantum mechanics. Lockwood (1996) and Mensky (2005, 2007a, 2010) have been suggesting the same in the context of different versions of the multiverse interpretation of quantum mechanics. In this book, this strategy has been advised within the last chapters, and the clustered-minds multiverse been suggested as an interpretation of the multiverse containing this feature. According to the top-down approach, *reality is largely*

[4]It is actually quite difficult to classify the position taken by Zeh. Zeh (1999, 83), as a many-minds theorist, is appreciative of the fact that creating an actual irreversibility is finally a matter of the interpretation of quantum mechanics: "The question for the origin of irreversibility is therefore intimately related to the *interpretation* of quantum theory (…)." [Italicizing by Zeh] It is also clear that—unlike many others—he is fully aware of the lessons to be taken from the timelessness of the Wheeler-DeWitt equation (see Zeh 2012, Chaps. 16 and 17). However, according to my—perhaps erroneous—reading of his publications, I felt that Zeh finally opts for an 'almost objective' status of irreversibility within the theory of decoherence leading to an 'almost objective' status of time, at least for all practical purposes (FAPP) and within our regular domain of life. This all seems to be related to the fact that he has a critical position with respect to Heisenberg's idealistic view on quantum mechanics and hence would not buy into the top-down-decoherence idea presented in box 2.1, earlier in this book and elsewhere. Again, as a many-minds theorist, Zeh does see part of the subjectivity of the emergence of our picture of the world. So, what probably characterizes Zeh's position with respect to time would be an 'almost objective' time arrow from irreversibility within an open-systems perspective, but within a many-minds view of quantum mechanics, and taking into account the fact in a closed-system perspective, time does not exist.

constructed, starting within consciousness. This comprises a *subjective* selection of the *preferred basis* (a mainly collective effort in the clustered-minds multiverse; see also Chaps. 6 and 13), a subjectively decided upon degree of consciousness allocated to different realities, a *subjective* flow of *time*, etc. Looking at all those phenomena as being subjective, the problems that have been encountered within the bottom-up framework disappear. E.g., if the preferred basis is selected subjectively, the plethora of possible bases encountered in bottom-up analyses might be narrowed down considerably in principle. This reasoning might rightly be considered as somewhat close to an *idealist* position. This first conjecture will further be analyzed and a classification specified in more detail, in the following.

Physicalism, Supervenience, and the 'Hard Problem' of Consciousness

Proponents of physicalism typically employ the argument of *supervenience* of the mental upon the physical (Davidson 1980 [1970]; Lewis 1994; for the definition of supervenience see also Hare 1984). This means that there cannot be a change in conscious experience without a change in the underlying physics (e.g., Davidson 1980 [1970]; Lewis 1994). Specifically, if (a) two events are identical in all physical respects, they cannot differ in any mental respect; and (b) if any mental aspect is supposed to be altered, some physical aspect has to be altered, too (Davidson 1980 [1970]). This way, consciousness becomes sort of a 'by-product' of the physical, perhaps even an *illusion* (Dennet 2003a). Please note the interesting correspondence with the other extreme sketched in the introduction of this chapter: monistic Indian idealism, calling the *physical* an illusion.

An important distinction has been made between strong and weak supervenience by Kim (1993) [1984]. Explaining this difference will sound as if Kim had already been discussing multiverse concepts; the critical distinction here is whether one or more world(s) is (are) concerned. But Kim did not actually discuss the multiverse. Kim's discussion is a purely hypothetical one, a thought experiment. In his theory, alternative worlds are a *metaphysical possibility*. Within Kim's theory, *weak supervenience* is identical to the basic concept of supervenience in the usage by others, including Davidson (1980) [1970] as well as the above definition. Specifically, it refers to one world w, where if two individuals x and y share a certain set of properties of some type, e.g., neurological, they must also share the properties of some other type, e.g., mental events, if mental properties are supposed to supervene on the neurological ones; whereas *strong supervenience* is defined based upon more than one (hypothetical) world (Kim 1993 [1984]). Here, keeping the example, if individual x resides in a world w_1 and individual y resides in a world w_2, it is still required that if the two individuals x and y share a certain set of neurological features, they also share the mental properties if the mental is supposed to supervene on the neurological.

In this chapter, I am not interested in supervenience of consciousness on the brain, but rather on the wave function—because this is the more basic level of analysis if the framework of quantum mechanics is to be applied (even if the brain might be seen as the link between consciousness and other parts of physical reality by many; eventually, the brain is part of the wave function, too). And since *strong* supervenience is the tougher requirement of the two, it is clear that strong supervenience has to be rejected for the case of the clustered-minds multiverse if physicalism is to be rejected. This is a typical philosophical strategy using the concept of supervenience.[5]

A concept by David Chalmers is a good starting point for the following analysis: his thought experiment on *zombies* (for Chalmers' exact definition of philosophical zombies see Chalmers 2010, 106–107). But let me start with briefly revisiting his earlier—and quite famous—argument, his 'hard-problem' analysis (Chalmers 1995; see also Chap. 3). Physicalism might be dominating the contemporary analysis of the mind-body problem. However, Chalmers (1995) 'hard-problem' concept has caught a lot of attention, too, partially because it tries to justify dualism. In his 'hard-problem' analysis of consciousness, Chalmers (1995) suggests that consciousness might be something 'special,' something that might not easily be reducible to the physical. For Chalmers, the hard problem of consciousness is the explanation of *qualia*, the subjective experience by the individual of things happening, of colors, odors etc. Somewhat earlier than Chalmers, Erwin Schrödinger (2004) [1958], in his well-known quote, foreshadowed this idea:

> The sensation of color cannot be accounted for by the physicist's objective picture of light-waves. Could the physiologist account for it, if he had fuller knowledge than he has of the processes in the retina and the nervous processes set up by them in the optical nerve bundles and in the brain? I do not think so. (154)

It should be noted that qualia are as much characterized by the experience of subjectivity as the measurement problem in quantum mechanics is characterized by the subjective perspective of the observer. Explaining consciousness and solving the measurement problem of quantum mechanics are obviously related (see for this thought, e.g., Mensky 2005, 2010; see also Fig. 3.1 as well as the explanation of this figure in Chap. 3).

Let me now turn to Chalmers' thought experiment on zombies (for an earlier related usage of the idea of philosophical zombies see Kripke 1980 [1972]). Simplifying, a philosophical zombie might be defined as a physically identical copy of a person on our planet; this copy is living on a different planet; however, *the zombie does not experience anything* since it lacks consciousness. Chalmers now metaphysically constructs an entire world full of such zombies, a world without qualia (as just defined; see also Chap. 3); and this entire world physically exactly resembles ours. Since such a world is conceivable, he argues, it is metaphysically possible and this argument can be used *against* physicalism since physicalism states that *all is physical*. But if all is physical, there should be no difference between the two worlds with and without consciousness. If there nevertheless is a difference, consciousness is not reducible to the physical.

---

[5]According to McLaughlin (1984, 1995), this style of argumentation might be called 'argument by appeal to a false implied supervenience thesis' (or argument by appeal to FIST).

It can be shown that in the course of the argument, Chalmers actually rejects strong supervenience. If individual x resides in world w_1 and this is the world with regular individuals (with consciousness); and if a physically exact replica of individual x, i.e., individual y, resides in a world w_2 and this is the world with zombies, this rejects the consequence that assuming the mental supervenes on the physical, the two should share the same mental properties since they share the same physical ones. Being a zombie is a mentally different state than being an individual with consciousness.

I would like to additionally remind the reader of the *teleological perspective* pursued in Chap. 3 leading to the following question: What is the *sense* of qualia? In Chap. 3, this question was answered via the existence of *free will* that is executed by consciousness (see also Chap. 6). But then, doesn't looking for 'functions' of qualia, e.g., in the form of performing free will or any other function such as humans' ability to reflect upon abstract concepts or their creativity, require consciousness even having an *influence* on the physical, basically reversing the above supervenience relation (see box 5.1 at the beginning of this chapter)? And is this a problem?

Specifically, the following *dilemma* might be constructed by a physicalist. This, however, requires the 'usual' assumption of a *singular universe*. Since in a singular universe, the degrees of freedom of interplay between the mind and the physical are 'somewhat limited,' the physicalist might find it easy to make the point that either consciousness must be supervenient on the physical (it then just being a by-product of the physical) or that consciousness should have an impact on the physical, leading to various objections such as violating the conservation of energy postulate or the closure principle (for a critical review and discussion of those objections see, however, Collins 2011, Goetz 2011, Kuczynski 2004). Interestingly, this dilemma loses its threat completely within the clustered-minds multiverse. This will be demonstrated in the following section.

Supervenience, Decoherence, and the Multiverse

Does consciousness supervene on the wave function in the clustered-minds multiverse? In order to reject strong supervenience, we actually need to look at *two universes*, both described by an identical Schrödinger wave function.[6] To be precise, this implies looking at *two multiverses*. Each *individual* is represented by an unknown number of versions.[7] Let me look at only one individual now. Given the type of *free will* specified in Chap. 4 (see also Chap. 6 and especially Chap. 8), consciousness can be allocated differently between the different versions of one

[6]Perhaps, the Wheeler-DeWitt equation (DeWitt 1967) would be more appropriate, here, since I am talking about an entire *cosmos I* and an entire *cosmos II*, but this does not change my basic reasoning. The trouble is that there currently is no 'world formula' generally accepted as a complete and accurate description of the cosmos.

[7]The rejection of strong supervenience is possible both on the level of the 'individual' and on the level of the 'version.'

individual in *multiverse I* rather than between the versions of the corresponding individual in *multiverse II*. For instance, whereas in *multiverse I* only the happy versions of the individual might be equipped with a high degree of consciousness, in *multiverse II* a lot of consciousness might reside with the unhappy versions.

There is a structural similarity between the metaphysical situation with two multiverses and Chalmers' metaphysical considerations regarding one *singular-reality* world *with qualia* and one *twin zombie world* that I have considered above. Chalmers' two worlds are physically identical, but the allocation of consciousness differs. My two multiverses are also physically identical in the sense that they are described by the same wave function. There are also differences between Chalmers' metaphysical situation and our metaphysical situation. But they are *not* relevant for the rejection of strong supervenience that applies to both. Specifically, one difference between Chalmers' and our example is that the allocation of consciousness to one of Chalmers' worlds is maximal, the other is minimal. Such extreme cases are impossible in the clustered-minds multiverse; they are excluded by definition (see Chaps. 4 and 8): The degree of consciousness allocated to some version of the individual, e.g., may never be zero. The other difference is that Chalmers compares two singular-reality worlds, whereas I am comparing two multiverses. Neither the extremeness of the allocation of consciousness nor the uniqueness of reality are of principal concern for the rejection of strong supervenience. It is clear that two individuals, or versions, for that matter, might differ in the mental state without the physical situation differing in any detail. Or in other words, as was possible based on Chalmers' (2010) analysis for the case of singular realities, I am now able to reject strong supervenience for the clustered-minds multiverse since the *mental situation* within the two physically identical multiverses might differ.

At the end of the last section I described a dilemma that might be constructed by a physicalist; and this dilemma is a real challenge within a *singular* reality: Either consciousness must supervene upon the physical, or consciousness should have an impact on the physical.[8] The latter is connected with all kinds of objections (see above). The first claim has just been dealt with. I was able to demonstrate that strong supervenience of consciousness on the physical can be *rejected* within the clustered-minds multiverse. Does that imply that consciousness must have an impact on the physical, and are the above objections (e.g., conservation of energy) to such an impact of consciousness on the physical relevant, then? The answer is actually quite simple.

I was able to reject strong supervenience via different allocations of consciousness between two multiverses. This alone, interestingly, does not imply an impact of consciousness on the physical. However, within a fully subjective interpretation of the quantum mechanical multiverse and following my definition of 'subjective physics' from box 5.1, the entire process of decoherence is supposed to start within

[8]Even the argument of psychophysical parallelism (see box 5.2 at the end of the chapter) would not be helpful within a singular reality since this postulate would be consistent with supervenience of consciousness on the physical. Thus, if supervenience is rejected, so is psychophysical parallelism.

consciousness. Within an open-system perspective, this *is* an influence on the physical. I.e., picking a certain preferred basis (or better: defining the structure of reality looked at) and inducing the decoherent histories that are following from there are actions of consciousness (some of them collective, cluster-based). Nevertheless, none of those changes occur on the level of the closed system of the entire multiverse. Hence, potential problems such as a violation of the conservation of energy postulate or closure postulate (see above) are not relevant.[9]

Within an open-system perspective and adopting the proposed subjective view, the wave function with its feature of entanglement is the way our 'playing field' is organized or set up (our collectively produced 'TV program' constructed). If we expect something 'material' and 'objective' out there, our experience turns out to be an *illusion*. What we do experience is part of the plethora of perceptual possibilities the wave function allows for. Note that this does not imply that we are alone, however: *It is a collective program.* Since the influence of other individuals is relevant for (co-)determining the preferred basis (a joint reality) and the decoherence processes following from there (decoherence I, i.e., entanglement, to be precise; see Chap. 2), and since the wave equation and its features (including the auxiliary Born rule for measurements along one reality) determine what we are able to experience, this is not indicating a purely idealist position, however.

Introducing the Concept of Dualistic Idealism

Towards the end of the last section, I have used the term 'illusion.' Indeed, one version of a respective individual, you or I for instance, experiences only one reality. Illusion, or *Maya* (a word from Sanskrit), is the term that both the *Vedas* and the *Upanishads* use for describing that our 'reality' is but a small and potentially misleading part of the larger truth. Indian idealism has been used as a term to describe the common theme of the various Indian philosophical traditions, greatly introduced and discussed in the monograph by Dasgupta (1962). On the one hand, the Indian philosophical concept of Maya is close to an accurate description of quantum reality in the form of the multiverse and Maya is often associated with Indian *monist* idealism. On the other hand, however, I would like to argue that a dualistic position is capturing the role of the wave function with its quite central law of entanglement and the Born rule as its auxiliary equation, constituting our 'playing field,' somewhat better. Therefore, monist idealism will not be proposed, here. The view proposed in this chapter is a dualistic view with a strong 'idealistic flavor.'

[9]An influence of consciousness on the physical has also been suggested within the collapse view (the so-called standard version of quantum mechanics and part of the umbrella of the Copenhagen interpretation of quantum mechanics) by Halvorson (2011) as well as Stapp (2009). Consciousness is here seen as responsible for collapsing the wave function.

Why might the term *dualistic idealism* be appropriate to describe the mind-body consequences of the clustered-minds multiverse proposed in this monograph? First of all, I would like to argue that consciousness (i.e., mind, split into versions of each individual, and including the minds of all other individuals) and the 'physical' (i.e., the wave function with its entanglement feature and the auxiliary Born equation) are still *different realms*. Specifically, consciousness has an influence on the realities to be perceived and makes the resulting experiences. But it *needs* the wave function and entanglement to experience it in the form we call *life*. Certainly, the multiverse is not a very pre-structured form of reality, but it becomes somewhat structured for an individual within 'consciousness clusters,' via the joint specification of the preferred basis etc. This is still a place with a real plethora of possibilities because of the multiverse feature, the fact that we might experience alternative realities. But the existence of a wave function, its linearity, the Born rule and entanglement are enough for me to hold that we are talking about a 'realm' in its own right. At the same time, the fact that consciousness of each version of an individual has only access to a small part of quantum reality, i.e., that the subjective perspective is so 'narrow,' that what we experience is so much dependent on the *emphasis* we put on different realities, that idealism is clearly the correct way of describing the situation. This is the reason for the term *dualistic idealism* coined here.

Box 5.2: Dualistic idealism or rather weak psychophysical parallelism?
Another way of looking at the problem would potentially be offered by the concept of psychophysical parallelism, discussed quite early, in different forms, by Gottfried Wilhelm Leibnitz (1898) [1714], Gustav Theodor Fechner (1860) and Wilhelm Wundt (1894), because quantum decoherence and processes in consciousness could be considered to be *somewhat parallel*. One would have to add 'weak,' to make it a *weak psychophysical parallelism*, because in the multiverse, there are many ways of experiencing the quantum world within consciousness, and I have argued (within Chap. 4) that different versions of the individual might have attached different degrees of consciousness to them. Moreover, some relationship between decoherence and the brain or other units of the physical body would have to be additionally postulated, or even better: empirically validated. But then, consciousness cannot actually *leave* the possibilities given by the wave equation and decoherence, so that some parallelism is kept. The reason why, although the concept of psychophysical parallelism is well-received among scientists, I did not opt for this concept for the core of this chapter but for dualistic idealism is simple. The idea within psychophysical parallelism is that things in physics and in consciousness happen in parallel, but either because of a perfect pre-harmonization by God (Leibnitz) or because of different perspectives on the same thing (Fechner) etc., but *not* because of a *causal* influence of one of the two on the other of the two. But there *is* a causal influence of consciousness on decoherence within the subjective, the open-system perspective. Dualistic idealism is simply the more appropriate concept since it allows for such influences.

Chapter 6
A Special Form of Free Will: Parallel Watching of Different 'Movies,' but with Different Levels of Awareness

Opening the Debate

The free-will problem has kept philosophers busy for more than two thousand years, and some disputes, such as the one "between compatibilists and incompatibilists[,] must be one of the most persistent and heated deadlocks in Western philosophy" (Nichols and Knobe 2007). It is hence impossible to provide an account of all the positions that have been taken so far; indeed, the history of the free-will problem in philosophy would easily fill a monograph. Similar to the discussion of the mind-body problem in the last chapter, this chapter will hence concern itself with a description and discussion of some typical positions that will later be reflected upon in light of the clustered-minds multiverse.

Let me start the discussion with the excellent definition of *compatibilism* from the *Stanford Encyclopedia of Philosophy*, since it contains all basic dimensions that constitute the ancient as well as current debate on free will in philosophy:

> Compatibilism offers a solution to the free will problem, which concerns a disputed incompatibility between free will and determinism. Compatibilism is the thesis that free will is compatible with determinism. Because free will is typically taken to be a necessary condition of moral responsibility, compatibilism is sometimes expressed as a thesis about the compatibility between moral responsibility and determinism. (McKenna and Coates 2015)

According to this definition (see Nichols and Knobe 2007, for an equally structured view on the problem), the first question is whether determinism is judged to be a correct description of the world; let me add that what the traditional discussion is concerned with is *determinism in a singular world*. Determinism in a singular world means that there is an inescapable course of action that *the* world or *the* history takes. The second question then is, *if* determinism (in a singular world) is true, whether this situation is *compatible with the existence of free will*. Let me note that if such determinism is rejected, however, it is still not clear whether free will exists. Some have, e.g., debated the free-will friendliness of an *indeterministic*

© Springer Nature Switzerland AG 2018
C. D. Schade, *Free Will and Consciousness in the Multiverse*,
https://doi.org/10.1007/978-3-030-03583-9_6

world. Whereas on the one hand, free will has been related very early to the seeming indeterminism of quantum mechanics as reflected, e.g., in the Born (1926) rule (see Lillie 1927), many have openly criticized the position that such an indeterminism might open the door for an actual free will (e.g., Walter 2001).

Coming back to the initial question of compatibility of free will with determinism, the compatibilist answers "Yes," and the incompatibilist answers "No." The third question, as could already be conjectured, based on the above Stanford Encyclopedia reference, is *whether moral responsibility holds* if determinism is true. This question can, in parallel to the question of free will, however, also be asked for the case of an indeterministic world. Whereas moral responsibility will—with the exception of a few overlapping thoughts by some of the referenced authors—separately be dealt with in Chap. 7, the first two questions (as well as the related alternative problems posed by indeterminism) are at the center of this chapter.

As can already be guessed from the state of affairs sketched within the last few paragraphs, whereas quantum mechanics (in singular-reality versions) partially has, the multiverse interpretation of quantum mechanics has not been integrated into the discussion on free will, so far (one exception is a very brief treatment within a box in Wallace 2012c, that is compatibilist in nature and has nothing in common with the treatment introduced here). And the important question to be addressed in this chapter is how the perspective on some of the philosophical issues might change if this is done explicitly; specifically, what will happen if the old disputes are analyzed in light of a free-will-friendly version of the multiverse, the clustered-minds multiverse?

To this means, this chapter will proceed as follows. In the next section, a set of (stereo-) typical, including the most extreme, positions on free will are sketched. The idea is to achieve an overview of the *scope* of traditional approaches in philosophy. The subsequent section will show how some of the disputes over extreme positions disappear when the problem is analyzed from the perspective of the clustered-minds multiverse. The section ends with Box 6.1 that concerns itself with a critical evaluation of an alternative explanation for free will, recently offered by Hameroff (2012), based on brain biology and the objective reduction modification of quantum mechanics. Finally, the chapter will concern itself in more detail with the type of free will we are granted with according to the clustered-minds multiverse. Specifically, this section will analyze whether we are talking about choices or rather perceptions when looking at the flow of experiences according to this approach; and it tries to further describe the kinds of perceptions relevant, here. This is supposed to address one of the criticisms that was formulated with respect to the EEC by Mensky (actually, its impreciseness in this regard); clearly, this criticism would also apply to the clustered-minds multiverse if not a novel solution would be suggested, here. The chapter ends with two more boxes. Box 6.2 discusses whether, in addition to the main route to free will, described in this chapter, the subjectively selected preferred basis and top-down decoherence might also constitute forms of free will. Box 6.3 contains a piece of theory development: From time to time, consciousness might be *reallocated*, literally *across* realities, to deal with (unexpectedly) negative developments within the experience/a 'history' of one version of the individual. This action enlarges free will, however, it challenges the idea of constant memories; the necessity of a 'quantum brainwash' is discussed.

Some Stereotypical Positions in the Free-Will Debate: Short Descriptions and Critical Evaluations

To start with, I would like to propose the following four positions that have been taken in the philosophical debate on free will as being stereotypical and hence most reflective of the state of affairs.[1] Whereas the first two positions hold that free will exists (albeit for different reasons), the two latter positions deny its existence (actually for related reasons):

(a) The libertarian perspective proposes that determinism is false and that free will exists.
(b) The compatibilist' perspective proposes that free will exists despite determinism being correct.
(c) The determinist-incompatibilist' perspective states that determinism is true and hence free will is absent.
(d) The indeterminist-non-libertarian perspective states that determinism is not true but free will is nevertheless absent.

Those four positions will underlie the following analysis and discussion. Positions (a) and (b) will be dealt with separately, positions (c) and (d) will be analyzed together.

(a) Libertarianism is rejecting determinism. Hence, the question of compatibility of free will with determinism [as discussed in (b)] becomes obsolete within this framework. The most well-known contemporary proponent of libertarianism is *Robert Hilary Kane* who developed some related versions of understanding free will and responsibility out of *quantum indeterminism* (e.g., Kane 1985, 2003). The basic idea is always that quantum indeterminism, a typical feature of quantum mechanics according to the Copenhagen interpretation (as umbrella

[1]There are several classifications out there of the many different positions taken in the free-will debate that, however, mostly use the terms "compatibilism," "incompatibilism," "libertarianism," "determinism" and "indeterminism." Exploring all logically possible combinations between those terms (or, alternatively, reporting at least on all those that have been explored by authors in philosophy) is not possible here. The labels used for positions (c) and (d) are not precisely identical with those mostly used in the literature.

term) or singular-reality modifications (see Chap. 2),[2] is amplified in the brain somehow to create *alternative possibilities*. In recent versions of his theory, he sees the entire indeterminacy across all parts of the brain as the source of free will. He does not claim that most (or even all) decisions are free, but that certain self-forming actions that later co-manifest the character of a person are possible (e.g., Kane 2015). Dennett (1978) surprisingly gave "the libertarians what they want" (part of the title of his contribution), by hypothesizing the existence of sort of a *chance stage* in the decision making by the brain preceding the selection stage, but this was clearly not what Kane wanted (Kane 2005). Kane, across different publications, does not associate free will with some random outcomes but rather argues that the indeterminism in the brain *can be used* to produce free will. The biggest criticism that I would tentatively raise with respect to his approach is, however, the *vagueness* as to how the possibility of free choices is supposed to emerge from indeterminate quantum processes.

(b) Compatibilism requires taking a position on free will that is *not* dependent on the *actual* existence of alternative possibilities (under the same given circumstances). A definition clearly *excluding* compatibilism is the following by Ginet (1997): "(…) an action is free if and only if up until the time of the action the agent had it open to her not to perform it: she could then have performed some other action instead or not acted at all" (85). If we are "(…) seldom if ever able to act otherwise than we actually do," (van Inwagen 1994, 95) the question is how a compatibilist perspective might be justified.[3] In a way, compatibilists associate free will with an *absence of outer limitations* or strong force; whilst the 'outer' might actually comprise the brain and other 'inner' processes in some but not all versions of compatibilism (see Walter 2001). According to David Hume (1967 [1740]), e.g., "this (…) [type of] liberty is universally allowed to belong to everyone who is not a prisoner and in chains." One of the most important contemporary compatibilists is Harry Frankfurt. He rejects the, as he calls it: *principle of alternative possibilities* (PAP) underlying most incompatibilist arguments such as the one quoted above by Ginet (1997): "(…) the principle of alternative possibilities is false. A person may well be morally responsible for what he has done even though he could not have done otherwise" (Frankfurt 1969, 828). The point he makes is based on thought experiments of the following type. He assumes, e.g., a hypothetical person named Jones who has made a decision already. Jones is then *forced* to act in a certain way by a *severe threat* imposed on him. The threat is supposed to be so

[2]Quantum indeterminism is mostly associated with the Copenhagen interpretation (and other collapse versions including modifications) of quantum mechanics because of the apparent randomness implied here by the Born (1926) rule. It can, however, also be routed in the Heisenberg (1927) uncertainty principle that is compatible, however, with the Copenhagen interpretation. But I personally do not see how the Heisenberg uncertainty principle *per se* would give rise to different possibilities in the brain.

[3]For a debate with respect to the definition of important terms relevant for the compatibilist position see the papers by Fischer and Ravizza (1992) and van Inwagen (1994).

substantial that most people would have obeyed. Frankfurt then considers different cases where the original decision is either in conflict or in line with the action enforced by the threat, and he also considers different versions of Jones being more or less affected by (or resistant to) the threat. And he then constructs situations where one would or would not hold Jones responsible for his action (Frankfurt 1969). Whereas those so-called *Frankfurt cases* are very appealing, intuitively, I would still defend the incompatibilist position, however. Frankfurt argues that the degree as to which we may want to hold Jones responsible for his actions depends on the roles played by his original decision and by the threat. From my perspective, however, Jones might not be held responsible even in the most extreme case where he had made the decision to, say, commit a crime, already, and would not have needed the threat at all to carry out that action. The reason is that in a (singular) deterministic universe, a person simply has no influence on either her character nor on the *mixture* of motives leading to certain decisions even in the absence of any threat.

(c) and (d) An especially crisp exposure of *incompatibilism* as well as *non-libertarianism, excluding* the possibility of *free will* for both deterministic and indeterministic universes, respectively, is the one by Broad (1952 [1934]). For Broad (as for Frankfurt) the question of free will is closely intertwined with the question of responsibility (for more details and different positions on the question of responsibility see Chap. 7) that he calls *obligability*. According to Broad, in order to hold someone responsible, the following must apply: "We must be able to say of an action, which was done, that it could have been avoided, in some sense of 'could' which is not definable in terms of 'would have, if.' And we must be able to say of a conceivable action, which was not done, that it could have been done, in some sense of 'could' which is not definable in terms of 'would have, if.'" In other words, if alternative actions can only be chosen if some *other conditions* exist prior to the decision, Broad would see this as incompatible with an action that an individual can be held responsible for. Or again in other words, Broad strongly relies on the concept of alternative possibilities that Frankfurt so vigorously attacked. He furthermore makes clear that he sees indeterminism as no escape because the agent's actions would still not be self-determined. Thus, no matter whether the universe is a deterministic or an indeterministic one, he sees the conditions for free choices and responsible actions as not given: "They [i.e., people defending the existence of free will and responsibility,] would like to say that the putting forth of a certain amount of effort in a certain direction at a certain time is completely determined, but is determined in a unique and peculiar way. It is literally determined by the agent or self (…) [.] I am fairly sure that this is the kind of proposition which people who profess to believe in Free Will want to believe. I have, of course, stated it with a regrettable crudity, of which they would be incapable. Now it seems to me clear that such a view is impossible." From my perspective, *Broad's position is inevitable in a singular universe*. It can neither be rejected by the Frankfurt cases nor by libertarianism in Kane's (too vague, from my perspective) version of it. As I see it, a way of justifying free will is the acceptance of the clustered-minds multiverse proposed in Chap. 4, and, as will be shown in Chap. 7, the conditions for responsibility are still somewhat delicate.

The Extreme Positions Disappear: Free Will from the Perspective of the Clustered-Minds Multiverse

Within the clustered-minds multiverse (as well as within most other multiverse interpretations), there is *only* the linear and *deterministic* Schrödinger equation (and, for most authors, including myself, the Born rule as an auxiliary equation, more or less explicitly admitted, sometimes derived from other principles, in my case limited to the usage within one history; see Chap. 4). However, since there are multiple versions of reality that an individual might perceive (i.e., live in) because of entanglement with different states (i.e., relative states) of the individual's consciousness (in turn separated into parts of consciousness allocated to different versions of the individual), the *prerequisite* of *alternative possibilities* is given. Also, as introduced in Chap. 4, free will is executed in the following way: Consciousness might decide to put more or less *emphasis* on different realities; or, as stated in the title of this chapter, it might decide in favor of more or less *awareness* with respect to different 'movies.'

To prepare the later section on "Choice or Perception …," dealing with the question as to how exactly the type of free will provided within the clustered-minds multiverse looks like, let me put the above a bit more figuratively now and let me furthermore assume—just for the sake of imagination—that the individual is able to *explicitly* choose between different 'movies,' seeing all of them at once. Then just imagine a huge wall with different screens where all those 'movies' are run in parallel, where in each of those movies a different version of the individual appears. Then the individual is free to choose which of those 'movies' to watch with what intensity. Even if it mainly watches only one or two of those movies, it will also see the adjoining ones fairly clearly, but it will hardly recognize anything of the story playing on the screens that are far away.[4] Let me examine what this as well as

[4]This picture uses the basic idea of the torchlight, the simplified one without the possibility of high degrees of consciousness in 'outer regions' (see Chap. 4), and it merges it with the idea of different 'movies.' Note that the almost omniscient individual assumed here, in principle having access to all parallel movies, is a fiction but might not be impossible within the multiverse framework—its actual existence depends on how consciousness is organized (the idea is similar to Mensky's ideas of superconsciousness and postcorrection, but without any recurrence on 'subjective probabilities;' see Mensky 2007b). Within dualistic idealism, experiences are limited by the possibilities provided within the wave equation. Without assuming strong psychophysical parallelism (see Chap. 5) such abilities of consciousness are conceivable; and my thoughts are definitely not inconsistent with quantum mechanics.

additional theoretical insights, based on the multiverse, imply with respect to the four above-described stereotypical positions in the philosophical debate on free will.

First of all, libertarian approaches based on *indeterminism* in the brain are not plausible[5]: Thus, whereas Kane's intuition with respect to quantum mechanics turned out to be correct: it is indeed the world described by quantum mechanics that forms the basis for free will, a different interpretation of this theory than that favored by Kane grants us with this precious property. Not the quantum indeterminism implied by the Copenhagen interpretation (and other collapse versions of quantum mechanics) but the plethora of worlds implied by the multiverse interpretation is able to solve the 'riddle' of free will. And since the brain is not 'playing roulette,' but consciousness is putting different emphasis on the realities to be experienced by different versions of the individual, the sometimes implicit, sometimes explicit criticism with respect to libertarian positions, such as the rather cynical critique formulated by Dennett "granting the libertarians with what they want …," can be addressed. *So, it is a new type of libertarianism that the clustered-minds multiverse grants us with.*

Second, compatibilist' positions become *less important* since the seeming tension between determinism and free will is diminished with respect to the free-will question (this, as already mentioned, turns out to be a bit more difficult with the issue of responsibility as will be shown in Chap. 7). 'Alternative possibilities' exist without violating determinism on the one hand and without adopting indeterminism on the other hand. Clearly, the reason is that the *type* of determinism that the wave function describes is a very different type of determinism than that which compatibilists had in mind. The classical notion of determinism is best reflected in the idea of mechanical clockwork, and that is indeed the picture that is often used as a figural expression of it. Determinism in a singular reality would lead to events at a certain time inevitably following from events in previous periods of time. But this is not the case in the clustered-minds multiverse. The wave function does not guarantee a quasi-classical reality on its own but has to be *interpreted* this way by (collective, i.e., minds-cluster-based) consciousness (preferred-basis problem; see Chap. 4). And consciousness not only (collectively) 'selects' quasi-classical frameworks, but it also (individually and freely) decides on the degree of consciousness (or more or less awareness) allocated to different realities.

(Deterministic) incompatibilism and (indeterministic) non-libertarianism [described in (c) and (d)] are both rejected. First, free will—in the special form just described—is not incompatible with determinism since the deterministic wave equation is granting us with the alternative possibilities required for free will as has already been pointed out. Second, (indeterministic) non-libertarianism is ruled out simply by the fact that there is no *actual* indeterminism out there.[6]

---

[5]Again, the Schrödinger equation is deterministic, the Born rule is only an auxiliary equation with a special purpose, and Kane's concept of indeterminism in the brain is somewhat vague, as has already been stated.

[6]The indeterminism of the Heisenberg (1927) uncertainty principle means something different. Partially, this is also a problem of the translation of a German term to English. The German term

In a way, with only a slight bending of terms, the clustered-minds multiverse implies a position of *libertarian compatibilism*, clearly a novel category. It is a libertarian position, because we have an *influence* on *what to experience* with *what degree of consciousness*; it is a compatibilist position, because although the wave equation is deterministic, we do possess free will. The type of free will granted within the clustered-minds multiverse may nevertheless be different from that many libertarian philosophers had in mind. This might have already been conjectured based on the subjectivity, the *dualistic idealism*, introduced in Chap. 5, that this concept implies; and this chapter adds to this perception. Things are complex, with respect to free will, when looking through the lens of the clustered-minds multiverse, but free will clearly exists.

Box 6.1: Is quantum brain biology able to save free will?[7]

Hameroff (2012) suggested a strategy of demonstrating the existence of free will that is similar, in some aspects, to the one that I have proposed in Chap. 3. Specifically, he argues that Libet-type findings might be consistent with free will if consciousness were able to influence the actions of the brain/body as well as individuals' choices[8] *'backwards'*; and, based on quantum brain biology, he is convinced that consciousness has this capability. However, Hameroff's theory, based on the objective reduction modification of quantum mechanics (Hameroff and Penrose 1995), is a singular-universe theory. Thus, paradoxes connected to time-backwards effects in a singular universe are relevant. Hameroff's argument that only 'acausal' information will be sent backwards and hence no paradoxes will materialize (Hameroff 2012, 11) cannot be accepted. Either the backwards-directed information changes *something* or it doesn't, where in the latter case it is simply *irrelevant*. Only the multiverse interpretation of quantum mechanics is able to account for 'changes' in the 'past' that are inspired by the 'future' and in turn change the 'future' (see Deutsch and Lockwood 1994). This is not about 'material changes' implemented in the 'physical world,' at least not in the multiverse (but it would have to be in a singular reality with all the consequences in terms of well-known paradoxes). The whole point is that there *are* those multiple realities where conscious emphasis might put different weights on, and that operating in a singular reality would not allow for *any* such backwards-directed changes.

for the Heisenberg principle is "Unschärferelation." A direct and certainly awkward translation of this term to English would be something like a "non-sharpness relation."

[7]See, for a preliminary analysis, Schade 2015, 351–352.

[8]As has already become clear within the last chapters and will be become even clearer within the next section of this chapter, talking about 'choices' is an extreme simplification.

Choice or Perception? A Detailed Look into the Type of Freedom Granted Within the Clustered-Minds Multiverse

Within this as well as in the previous chapters, I was able to establish the possibility of an actual free will in the sense of an existence of *alternative possibilities* and an influence of the individual on *what will be experienced with how much intensity or awareness*. It also became clear that the type of free will that the multiverse grants us with is different from the type of free will most libertarian (or other) philosophers might originally have had in mind: a *choice* between different, *objective* courses of reality.[9] Although I have casually touched this question a few times, already, it is now necessary to answer the question as to how that novel type of freedom exactly works.

Whilst dealing with one of the multiverse interpretations in Chap. 4, the EEC by Mensky (2005, 2007a, b, 2010), I have criticized Mensky—among other aspects—for *not providing a precise definition of free will that assigns clear roles to the concepts of perception and choice*.[10] In my concept of the clustered-minds multiverse, consciousness is able to put *different weights on different realities*. This seems still not to address, so far, that free will is normally associated with *making decisions*.

The course of action I will be taking here to provide clarity is to analyze a simple, practical (albeit stylized) decision problem within two scenarios: a standard decision-theoretic framework and the appropriate analysis within the multiverse (see, for a similar analysis, Schade 2015, 352–355; I have left out one of the scenarios presented there and added more theoretical background to the analysis of the remaining two). In the practical example, Julia wants to buy either a Mercedes or a Tesla. And the only criterion she is interested in is the *reliability* of the car (unlike in the example presented in Chap. 4 where Louise and Tim made a decision as a couple, no potential conflict between partners may arise, here; we may think of Julia being single).

Scenario 1: According to standard (non-quantum) decision theory, Julia wants to choose the alternative that maximizes her *expected utility* (for more details on this approach see Chap. 8). For each alternative, this is the utility of potential outcomes weighted by the probability of their occurrence. Hence, Julia will form her expectations with respect to the reliability of the two cars. She will determine the utility of different potential outcomes (e.g., number and costs of garage visits), taking into account her *risk propensity*[11] and will attach the respective *subjective probabilities* to those outcomes. Note that although these probabilities are indeed

---

[9]In most cases, such a view of reality is implicit. And I certainly do not want to argue that all philosophers talking about free will are non-idealists.

[10]In psychology, anyhow, perception and choice are traditionally treated as *different* processes (see, e.g., the textbooks by Hayes 1994; Lefton 1994).

[11]If Julia were risk averse, her utility function would be concave, if she were risk taking it would be convex (see Chap. 8).

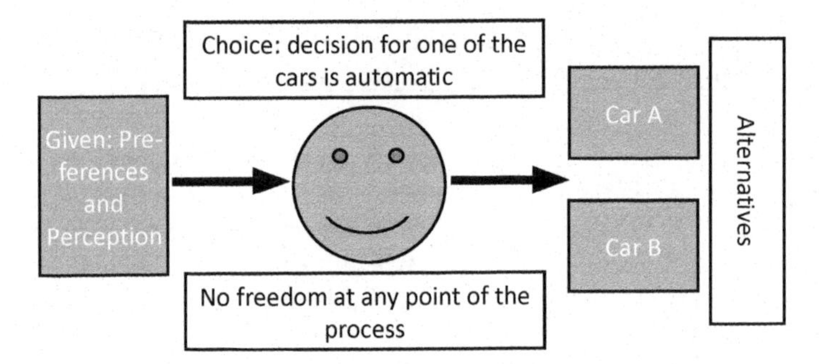

Fig. 6.1 Free will in a simplified decision-theoretic framework. *Source* Schade 2015, 353

seen as subjective (Savage 1954), Julia is kind of a 'robot' processing the information out there to generate those probabilities. She might come up with different probabilities than somebody else (which is allowed in the subjective expected utility framework), but she is assumed to try to come up with the best possible estimate. If she does not come up with the best estimate from her perspective, this would be considered an *error*, not free will. Since I have assumed only one dimension to matter, i.e., reliability, there are no, potentially cumbersome, tradeoffs between attributes.[12] Finally, we look at the simplest case here since there are only two alternatives.

Consequently, if Julia perceives the Mercedes as offering a higher expected utility than the Tesla, she will buy this car. If the Tesla offers the higher expected utility than the Mercedes, she will instead buy that car. Given her *preferences* and her *perceptions, her choice is fully determined.* This simplified decision-theoretic analysis is depicted in Fig. 6.1. In this as well as in the subsequent Fig. 6.2, the smiley represents the point where people *think* they decide.

Scenario 2: The appropriate analysis in the clustered-minds multiverse looks like this: Julia compares the *'attractiveness' of different alternative realities already including the choice of a specific car* (see Fig. 6.2). Part of those 'movies' are the different choices that Julia *makes* and *has made*. Julia can 'opt' between experiencing, with larger awareness, a reality (or in other words: allocate more consciousness to that reality) in which she first chooses a Mercedes and later has chosen a Mercedes (with the respective consequences) and a reality in which she first chooses a Tesla and later has chosen a Tesla (with the respective consequences). If the 'movie' with the Mercedes turns out to be more attractive (still with

[12]Using the word 'cumbersome,' I want to make the reader aware of the fact that one would either have to somehow integrate a multidimensional-utility perspective (see Chap. 9); an integration with the expected utility framework, however, is not trivial. Or one would have to use a utility function that takes the potential tradeoffs between those attributes into account (the latter would be consistent with a unidimensional utility function where the solution of such complexities is just assumed).

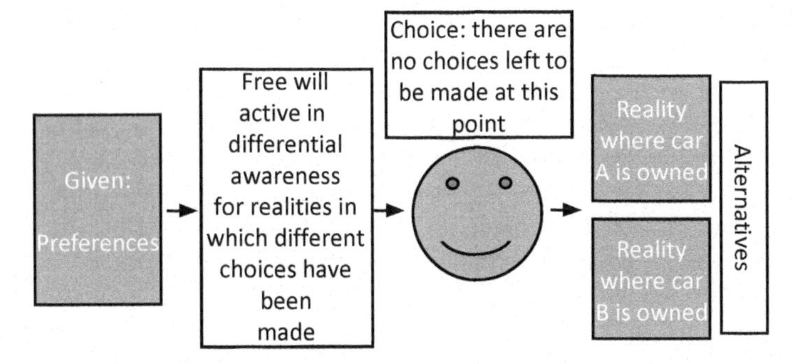

Fig. 6.2 Free will in allocation of awareness to different realities. *Source* Schade 2015, 354, modified

the reliability of the car being the only component that differs!), her consciousness will opt for allocating more consciousness to *this* reality.

It is important to realize how consistent this description is with what was discussed in Chap. 3 as a *reinterpretation* of the findings by Libet and coauthors. It was argued that consciousness might be able to work 'backwards.' The reality where brain areas for, say, moving the left instead of the right hand are activated would then 'ex post' be equipped with more awareness when consciousness has explored the different 'movies' associated with moving the right hand versus the left hand.[13]

Let me add that actual 'choices' between courses of action, in the sense of *altering* physical reality (scenario 1, Fig. 6.1) would imply leaving the linear development of the wave equation and thus its superposition principle—excluded within the multiverse interpretation of quantum mechanics. And 'choices' between experiencing different alternative realities, in the sense of a *hard selection* in the domain of consciousness, would lead to serious problems in terms of ontological plausibility (please revisit Chap. 4, for a discussion of the problem that Squires' first proposal encounters). Thus: *Within the clustered-minds multiverse, there are no 'actual choices.' We are instead allocating different degrees of consciousness (or different levels of awareness) to different 'movies' that contain versions of ours making different decisions and experiencing their consequences.* Thus, a certain decision is really just part of a respective movie, it *indicates* that our consciousness resides with a specific reality with that version of ours we just experience (sort of a

[13]One might see some similarities here with Mensky's (2007b) concept of *postcorrection*. However, it is actually hard to tell how similar the basic ideas are because Mensky does not discriminate between perception and choice. Also, a major difference between the two concepts is that Mensky supposes observation probabilities to be changed—an idea that was dismissed in Chap. 4 for a lack of theoretical consistency with the basic Schrödinger equation and ontological implausibility—, whereas here it is assumed that consciousness is putting different emphasis on different realities.

self-location information, if you will), but the actual *choices* within those movies per se have no physical consequence, anymore.

Having said all this, it is now also clear how we should describe the type of free will the clustered-minds multiverse grants us with. *We have the freedom of allocating more or less consciousness to realities where we are experiencing ourselves making one or the other choice.* The astonishing aspect here, most certainly, is that our intuition as to what a free choice is turns out to be violated. But this intuition is violated even more within deterministic, singular-universe theories declaring free choices an illusion.

Box 6.2: Free will from top-down decoherence and subjective selection of the preferred basis?

It became clear that substantial freedom can be generated from allocating more or less consciousness to different realities—where we have made different decisions and experience different courses of life. But aren't there more 'loopholes' for the execution of free will within the clustered-minds multiverse with its pronounced subjectivity? I would tend to disagree with respect to *individual freedom*. The perceptual freedom granted, in principle, via top-down decoherence and subjective selection of a preferred basis is mainly playing out within consciousness clusters. We, as a 'minds collective,' experience a certain 'time,' certain technology (e.g., digitalization), and a certain political situation, not just you or I. And sure, we *can* leave the framework, e.g., that of the preferred basis 'agreed upon' within our cluster, mainly via certain drugs or experiencing severe psychological problems. Some individuals are indeed experiencing odd, non-classical realities (e.g., within schizophrenia). But this type of freedom is dangerous, to say the least, it is not the type of freedom I have in mind, here. This thought will briefly be picked up again in Chap. 13. Otherwise, this book will not concern itself with those kinds of free choices.

Box 6.3: Reallocation of consciousness across realities and 'quantum brainwash'

Is consciousness supposed to be 'stuck' with the subjective 'history' of the version it has been allocated to? Is that even conceivable from a theoretical perspective? Consciousness is split into measurement outcomes with each measurement/decision; therefore the different versions of the individual, attached to those different measurement outcomes, would get less and less consciousness allocated to them—consciousness would get more and more *diluted* 'over time.'[14] Indeed, assuming that the total amount of consciousness an individual possesses (over all his versions) is *finite* (for a formal expression, on the level of an individual decision, see formula (1) in Chap. 8), it is absolutely clear that *reallocations* of consciousness across realities must take place from time to time. This might not be considered a *new type* of free will, but rather an ex-post correction[15] to the execution of free will in a dynamic perspective.[16] One might ask as to how those reallocations could be implemented and experienced (unconsciously?) and how this matches with our idea of free will, however. Chap. 13 will contain a quick discussion of this question, but a conclusive account will not be offered in this book. But let me, nevertheless, consider reallocations of consciousness an aspect of free will; they are, anyway quite similar to vectorial choices, just that they are not necessarily attached to actual decisions taking place. This is (a) good news with respect to free will that is now able to 'correct' 'past' allocations of consciousness, to react to unforeseen developments, but it leads to (b) 'odd news' with respect to memory. Let me, for the sake of simplicity, start with an extreme example with respect to (a): If a reality containing a lot of consciousness by the individual turns out to be unexpectedly unpleasant, consciousness might almost entirely (down to the minimum[17]) be allocated away from that reality, because the individual decides to better have that part of consciousness reside elsewhere.[18] The odd side of all this (b) becomes

[14]This is not to indirectly re-introduce the notion of a flow of time. The same basic idea pursued within the rest of this book: 'times as special cases of other universes,' is supposed to hold here. But subjectively, and that is the major issue dealt with in this box, versions of individuals experience themselves being in a specific 'history,' and the dilution of consciousness would be a subjective experience over subjective time.

[15]This is not meant to resemble Mensky's (2007b) idea of postcorrection; although, looking at my proposal only superficially, there seem to be some similarities with this idea. The reasons as to why there are differences between the two concepts might get clearer within the next footnote.

[16]One might ask as to why such corrections are even necessary—except for the dilution of consciousness problem. According to Mensky's idea of superconsciousness and postcorrection, individuals have access to the entire quantum reality and are able to optimize, taking all that knowledge into account. So, in a narrow sense, they should not even get to the point where they are directly facing a very negative development, unavoidably, say, with a highly conscious version of theirs. This makes only sense with (partially) myopic decision makers. Specifically, even though I do think that partial foresight is possible, mainly via the body (revisit Chap. 3 for the considerations on physiological anticipation), I do not, however, assume perfect foresight of an individual into all possible developments.

[17]I have assumed (in Chap. 4) that marginal consciousness remains with each reality.

[18]As in previous chapters, I am not able to say how exactly consciousness is organized and how the individual might be able to do that. Again, those thoughts might sound speculative, but they are not in contradiction with quantum mechanics. On a more practical note, I do not think that it will normally work this way; consciousness might not just be reduced from a high level to a marginal level within split seconds but smoothly allocated away from a certain version of the individual over 'time,' and only to a certain extent.

transparent if one makes this example even more extreme: What happens if consciousness is almost entirely taken away from a branch of related histories (because they all appear to end up in problematic situations, say)? A large chunk of consciousness is now allocated to realities where the respective versions of the individual experienced a completely different 'history,' e.g., have studied a different subject and are working in a different job. Does this, in turn, not imply that those parts of consciousness have to be 'quantum brainwashed' before they can be reintegrated with those parts of consciousness already residing with the respective other versions of the individual? Or is this problem perhaps entirely solved via the 'transferred' part of consciousness ending the attachment with the brain (and its encoded memories) of those versions that it leaves and gets attached with the brain of the other versions, possessing a different memory? But still, we make our experiences within consciousness, and memories appear to be highly relevant to our self-concept. This is puzzling. Taking our conscious experience as the natural reference point, this all seems to imply that memory is a difficult concept in the multiverse, somewhat malleable, so say the least. But somehow, this problem *is* solved, since we are not *experiencing* any discontinuities or even larger changes in our memories. A great article dealing with the problem of 'history,' of saying something meaningful about the 'past,' within the framework of multiverse quantum mechanics, even without taking all the problems of reallocating consciousness, encountered in this box, into account, is the publication by Squires (1992).

Chapter 7
Are We Responsible for Our Decisions?

Continuing the Debate

The last chapter concerned itself with the free-will problem from the perspective of stereotypical philosophical approaches as well as from the perspective of the clustered-minds multiverse. It became evident that the problem of responsibility is often seen as closely intertwined with the problem of free will so that the scholars that were referenced in that chapter were sometimes dealing with the two problems in parallel. However, the last chapter focused on the free-will part and left the responsibility part mostly to be addressed in the current chapter. There are good reasons for separating the analysis of the two issues within the context of this book. Whereas the answer to the free-will question is unambiguous from the perspective of the clustered-minds multiverse, responsibility turns out to be a more difficult subject.

In general, philosophical authors writing on free will, almost irrespective of the *kind* of free will they were looking at, felt that individuals can only be held responsible for their actions if that type of free will they defined as *relevant* within their theory was *provided*. An extreme case was made by incompatibilist philosopher Broad (1952 [1934]) who stated that holding people responsible for their actions is neither justified in a deterministic nor in an indeterministic universe since free will does not exist in any of them. What is the situation in the clustered-minds multiverse? The wave function is deterministic, but an individual has an *influence* on the level of awareness pertaining to different realities that are experienced by different versions of it. It is, however, not able to *perfectly* allocate consciousness to one and perfectly away from all other realities. This might be problematic, with respect to responsibility, in some cases: An individual, say, sitting in front of a judge might be a low-consciousness version of an individual that put most consciousness into a reality where no crime was committed. But, nevertheless, the person *is* sitting in front of the judge and has committed the crime in this reality. Does this imply that responsibility cannot easily be justified despite an

© Springer Nature Switzerland AG 2018

C. D. Schade, *Free Will and Consciousness in the Multiverse*,
https://doi.org/10.1007/978-3-030-03583-9_7

existence of free will? (Note that the possibility of reallocating consciousness, explained in Box 6.3 in the last chapter, is irrelevant for this matter.)

There are further potential caveats to claiming individuals fully responsible. One lies in the fact that many decisions we make are happening under a serious impact of *situational factors*; but individuals are often not aware of this impact. The question then is to what extent individuals should be held responsible for decisions under those conditions (Herdova 2016). The second caveat—but also a chance for responsibility as it will turn out!—is the fact that I have assumed (in Chap. 4) that individuals (also unconsciously, I suppose) avoid very-low-consciousness situations. So that it is difficult to impossible for them to leave the 'mainstream' of a minds cluster.

This chapter is organized as follows. In the next section, I will go more deeply into the philosophical debate on responsibility and check how the theoretical requirements specified within different approaches match with the situation within the clustered-minds multiverse. In the subsequent section, I am going to discuss how the impact of the unconscious on decisions plays out in terms of responsibility. I will then turn to the limits and chances resulting for the responsibility debate from the limits imposed on free will if people in fact avoid very-low-consciousness situations. Box 7.1 discusses a pragmatic, economic approach to 'punishment,' somehow pragmatically circumventing the necessity of holding people responsible for all of their choices for the justification of punishment of committed crimes. I will finally summarize the findings and offer a conclusion.

Who Is Morally Responsible Under Which Conditions? And Do Those Conditions in Principle Hold Within the Clustered-Minds Multiverse?

The definition of moral responsibility within western philosophy can be stated as follows: "(…) to be morally responsible for something, say an action, is to be worthy of a particular kind of reaction—praise, blame, or something akin to these—for having performed it" (Eshleman 2014). Philosophical thinking about responsibility is not only related to, but also about as old and diverse as the thinking about free will. Thus, I will use the same strategy as in the previous two chapters; I will only report on positions that I classify as representative for the debate without making any claim to be complete. Furthermore, on the one hand I do not want to be too repetitive, here, with respect to the discussion in the last chapter, and on the other hand there are important new insights and nuances that arise from the debate on responsibility that cannot directly be derived from the discussion on free will. The following discussion aims at a compromise in this regard.[1]

[1]To avoid any confusion, I will not deal with the difference between a merit-based view and a consequentialist view of responsibility, here. I will only deal with the corresponding theoretical

An important thought that has only partial correspondence with thoughts within the free-will debate stems from Aristotle's *Nicomachean Ethics*. In this work, Aristotle (~350 B.C. [1985], 1110a-1111b4) discriminates between two prerequisites of holding a person responsible for her actions: the action must have its *origin in the agent* and the agent must be *aware of what kind of action* he is taking and/or what he is *generating* with this action (see also Eshleman 2014). Whereas the first prerequisite is somewhat close to the question of an existence of free will in one form or another (at least would the complete denial of actions originating from within the agent be seen as incompatible not only with responsibility but also with any form of free will), the second and third are novel and also important for our below discussion of unconsciously driven action. With respect to the second and third requirements, it is clear that the required *awareness* might *only* be warranted with conscious action and with a clear view on *consequences* of one's own actions. Those two requirements turn out to be important when responsibility is analyzed within the framework of the clustered-minds multiverse. I will deal with them in the next section.

If Aristotle's first requirement of responsibility is applied to the clustered-minds multiverse, however, the above problem of an individual sitting in front of a judge can directly be rephrased as a problem of certain versions of the individual being only *imperfectly* described by actions originating from within the agent. Although the individual possesses the freedom of allocating only little consciousness to certain realities and consciousness might be highly concentrated in a few other realities, it is nevertheless impossible to *completely* remove it from the non-preferred ones. What are the consequences?

A well-known last-century philosopher that has concerned himself intensively with the issue of responsibility is Peter F. Strawson. In his seminal publication on "*Freedom and Resentment*" (1993 [1962]) he presents a specific version of a compatibilist account of responsibility. He uses strong, provocative, somewhat 'anti-theoretical' arguments such as the following: "If I am asked which (…) [party] I belong to, I must say it is the (…) party of those who do not know what the thesis of determinism is" (45). The type of approach that Strawson is advocating can be derived from the following passage on psychiatric patients:

> (…) [An] important subgroup of cases allows that the circumstances were normal, but presents the agent as psychologically abnormal – or as morally undeveloped. The agent was himself; but he is warped or deranged, neurotic or just a child. When we see someone in such a light as this, all our reactive attitudes tend to be profoundly modified. (Strawson 1993 [1962], 52)

Many—including myself—would emotionally follow Strawson's above thoughts. But as this quote correctly suggests, Strawson—in this regard preceding Frankfurt (1969)—develops a *case-based account* of responsibility. Such a

concepts, i.e. incompatibilism, often connected with the merit-based view, and compatibilism, often connected with the consequentialist view. For this distinction and further literature see Eshleman (2014). The exception I am going to deal with, the approach by Strawson, might be seen as a merit-based form of compatibilism (Eshleman 2014).

case-based account lacks, at least from my perspective, a strong appeal for usage within conceptual discussions. I would thus like to argue that although the large impact of Strawson's contributions on philosophical thinking with respect to the responsibility question in the second half of the last century is unquestionable, I do not feel more convinced of compatibilist' arguments with respect to moral responsibility than I have been with respect to the closely related free-will problem in the last chapter.

It is interesting to note that my anti-compatibilist conjecture is consistent with folk intuitions as long as people are in an *unemotional* state. When asked, within questionnaire experiments, for an *abstract* compatibility of a deterministic universe with moral responsibility (Nichols and Knobe 2007), a large majority of people would deny that somebody in a deterministic universe can be held responsible for what he is doing. However, the situation changes when *emotional* situations are presented in those questionnaire experiments such as violent crime examples: "(…) when asked questions that trigger emotions, (…) [peoples'] answers become far more compatibilist" (Nichols and Knobe 2007, 664). Or in other words, if someone committed an awful crime most people want to hold him responsible even in a *deterministic* universe. It is important to see, at this point, that this is exactly the situation we are oftentimes confronted with in jurisprudence so that Strawson's as well as Frankfurt's case-based compatibilist reasoning somehow matches with the issue most people would apply it to: *the question of moral responsibility in situations where people have broken the law*. It is at minimum unclear, however, how large the role played by emotions of people judging what somebody did *should* be in this context.

The question of responsibility within the clustered-minds multiverse does not easily match with the approach by Strawson just presented, not even in a pragmatic sense. Strawson appears to be uninterested in conceptual discussions such as those around the issue of determinism. But would Strawson still hold an individual fully responsible for her actions if there were only an imperfect coupling between that individual's preferences and the *observed behavior of one version of the individual* —the situation in the clustered-minds multiverse? Note that I have, within point 5 in the list of premises of the multiverse, stated that extreme allocations of consciousness are typically avoided (or have to be avoided). Thus, the idea of simply allocating the entire consciousness to the most ethical alternative, leaving all other realities with only marginal consciousness, and *requiring* such behavior from any moral decision maker, would be completely misleading.[2]

---

[2]A closely related limit to responsibility would arise from 'pre-specified' degrees of consciousness, based on mod squared amplitudes, a thought that was presented (and dismissed) in Chap. 4, within the implementation of the clustered-minds multiverse. Specifically, if not only the coupling between preferences and the degree of consciousness allocated to different realities would be indirect, but if free will could only partially adjust the pre-specified degrees of consciousness, then due to the sum of the two effects, responsibility would be even harder to justify. If one meets a version of an individual that, say, just committed a crime, how could we be sure that (a) this was not a reality with a large mod squared amplitude and (b) free will did just everything it could to reduce the degree of consciousness allocated to this reality?

Perhaps helpful in this regard are the more thought-provoking perspectives on responsibility that can be found in the eastern philosophical traditions. An especially concise summary of the Buddhist position has been published in an article by Abhayawansa (2013), as part of an ethical discourse (italics mine):

> In the context of moral behavior Buddhism uses the term *kusala* (wholesome) for the morally good conduct and *akusala* (unwholesome) for the immoral conduct. The judgment as to whether it is a morally good or bad conduct is done not on the basis of reward or punishment but on the one's *spiritual purity or impurity* and the nature of the (…) [effect] on *other people*. In this sense according to Buddhism, morality has both personal and social dimensions. Therefore, all the moral teachings of Buddhism have their significance not only to the doer but also to the society. This is further evident from the criteria of the moral concepts given by the Buddha in the Ambalatthika rahulovada-sutta in the Majjhima-nikaya. Here the Buddha tells Rahula that just as a mirror is meant for reflection; even so *every volitional act should be committed after proper reflection*. (…) [The individual] has to reflect on the consequences of the act that he is going to do, on himself and on others. If the act results in harm to oneself, to others and to both it should be reckoned as morally unwholesome. If the act brings about beneficial consequences to oneself, to others and to both it should be evaluated as morally wholesome. This tells us in Buddhism morality means *only the intentional activities of the man*, not the activities done without intention under the influence of any kind of authority of injunction or commandment.

What can be learned from this passage? Let me put the lessons in my own words and discuss whether they can be accommodated for within the clustered-minds multiverse: First of all, in Buddhism, the question of responsibility is *decoupled* from the question of reward or punishment. This may or may not make sense in a Western culture with its complex and highly developed law system (but see Box 7.1 on Gary S. Becker's approach, circumventing the necessity of holding someone responsible for his actions). Second, responsibility is a *personal* issue, closely related to the *development of the individual*. This is a very important thought. The incomplete coupling of preferences and consciousness allocated to different versions of the individual might increase or decrease, when a multi-decision path an individual is taking (over 'time') is considered within the clustered-minds multiverse. The reason is cluster membership together with the avoidance of very-low-consciousness situations (for more details with respect to that type of reasoning, see below). E.g., a person considering to steal a car might already be in a situation where such an action is *pre-committed*, somehow, partially by previous decisions. The idea is related to Kane's (2015) concept of *self-forming actions*. As has already been stated in the last chapter, Kane does not claim that most (or even all) decisions are free; the problem is similar, with respect to responsibility, to a situation where the impact of free will on which reality to enter is somewhat *incomplete*. In both cases, certain (perhaps multiple) self-forming actions can be thought of as later co-manifesting the character of a person (e.g., Kane 2015). And the character of the person might be seen as partially pre-committing certain types of decisions, e.g., via a membership in different minds clusters. This, in turn, might be seen as allowing for *some* responsibility of a person for her actions taken; since a person that permanently makes the 'wrong' decisions might not just be 'unlucky' ending up in problematic situations, but entered a minds cluster where committing a crime is one of the possible developments and will hence be realized in some reality—likely, but not necessarily, with a low degree of consciousness allocated to it. This type of responsibility is enhanced by the presented reallocation-of-consciousness possibilities presented in Box 6.3 in the last chapter. Specifically, whereas the relatively extreme example presented there for the sake of clarity—almost fully removing consciousness from one reality and distributing it elsewhere in one step—is very implausible, practically (see also the respective footnote attached to this example), it *is* plausible to assume that even subtle, long-term reallocations of consciousness would contribute to cluster membership and, thus, to the starting position for the 'next' decisions of the individual.

Third, responsibility is a *social* issue, connected to the impact that the deeds of a person have on society, whether they are harmful or beneficial to others. Within the clustered-minds multiverse, this is somewhat in tune, again, with the fact that realities of individuals *overlap*, partially because very-low-consciousness situations are avoided and because of meaningful clustering (see Chap. 10). Fourth, a person is obliged to *intensively reflect* upon her actions before acting, considering the effects on her and others. In a non-trivial sense, reflection can only have an impact on choices if there are alternative possibilities, a prerequisite that only holds in the multiverse. And fifth, *intention* matters. Similar to the considerations within the Frankfurt cases (see Chap. 6), a person who is *forced* into some behavior, might not be held responsible for it. The multiverse is neutral towards the integration of external force. However, as has just been pointed out with a reference to Kane (2015), in a long-term perspective self-forming actions might have an impact on the character of a person and hence on the intentions of an individual.

Especially unfamiliar for a western thinker, I suppose, is the situation in Confucianism:

> (…) not only is there no philosophical debate over the issue of freedom and moral responsibility in Confucian thought, there is not even a philosophical account of moral responsibility in Confucian ethical theory. (Hansen 1972, 169)

In his compelling comparative analysis between western and eastern philosophical concepts of responsibility, Hansen (1972, 171) points out that the, simplifying, two major ingredients in western philosophical theory of responsibility, 'moral codes' and 'excuse conditions,' as he calls them, do both not exist in Confucian thought. Specifically, whereas western ethics have a "structural affinity with law" (Hansen 1972, 172), no such affinity can be found in Confucianism. The existence of the *li* in Confucianism, norms of social behavior, could rather be explained as a tool used on the way to self-cultivation of a person (Hansen 1972, 177): "(…) they are a close approximation of how we would behave if we effortlessly and naturally, without reference to the code, followed our fully cultivated 'evaluating mind'" (ibid.). Furthermore, one could "(…) propose the following as a Confucian theory of moral responsibility. There is one rule: promote cultivation of character in yourself and others related to you in specific ways" (Hansen 1972, 184).

The Confucian view seems to match quite well with the situation in the clustered-minds multiverse. Long-term, self-forming actions—à la Kane—might 'move' the core of consciousness of an individual more and more towards situations where most outcomes of his choices are, anyway, 'positive' and he might be entering some of those situations with a high degree of consciousness, too. As has already been pointed out, this is close to Buddhist teaching, too. According to Buddhist thought, responsibility is something *personal*, directly related to an individual's state of development and with a close relationship to his intentions; Confucianism is just more direct and almost puts a 'must' on *self-cultivation*, makes *this* the overarching moral responsibility of an individual. Given the importance of this aspect in both eastern philosophies analyzed, it is interesting to note that the fact that consciousness has an impact on the reality to be experienced with what intensity, an important aspect of the clustered-minds multiverse, appears to have been foreshadowed by eastern traditions a long time ago. Also, in a more comprehensive perspective, self-cultivation is what makes us experiencing, with a higher degree of consciousness, 'better' versions of ourselves and hence 'better' realities (see Schade-Strohm 2017).[3]

Let me summarize. On the one hand, the clustered-mind multiverse is able to account for some complex ideas on (long-term) responsibility in *eastern* philosophical traditions. On the other hand, the idea that all (short-term) actions observed with some version of an individual unambiguously 'originate from within the agent,' a basic requirement for responsibility by Aristotle, is not fully granted because of the somewhat imperfect coupling of preferences with the situation certain lower-consciousness versions of the individual find themselves in. Three open questions are to be addressed in the following two sections of this chapter: (a) the importance of the unconscious for responsibility of an individual, (b) the importance of the avoidance of very-low-consciousness situations (together with the clustering of minds) for the responsibility of an individual, and, given the somewhat 'diluted' situation for (short-term) individual responsibility, overall, (c) a perhaps radical approach to justify a law system without recurrence to the strength of responsibility of an individual within a certain situation, dealt with in Box 7.1.

The Power of the Unconscious: A Caveat to Responsibility?

As already mentioned, Aristotle ($\sim$350 B.C. [1985], 1110a-1111b4) stated that not only must the action have its *origin in the agent*, but that the agent must also be *aware of what kind of action* he is taking and/or what he is *generating* with his action. The first (origin in the agent) has just been dealt with in some detail. The latter two are important further qualifications of responsible action. What would those exactly imply? Requiring us to know the consequences of our actions is a

[3]To be precise, in Schade-Strohm (2017), self-cultivation in connection with purposeful consciousness development makes you experience better realities (i.e., with more awareness).

tough requirement. In a narrow view, it is probably never fulfilled. If we loosen the requirement at bit, however, those two requirements together are actually close to the overall requirement that our *choices are consciously made*.

An illustrative textbook example in psychology that helps clarifying this matter would be learning to drive a car. Let me make the entire situation a bit more precarious by assuming that this is a car without automatic transmission. The beginner now wants to get the car moving for the first time (assume that the engine is already started) and will tend to have his consciousness closely monitoring the following actions: pushing down the clutch pedal with the left foot, using the right hand to move the gear shift to select the first gear, slowly releasing the clutch pedal, holding (too) tight the steering wheel etc. An experienced car driver would not pay much attention to all this and will do all these things unconsciously or *automatically*.

More recent approaches in psychology will not only state that very few processes are (fully) conscious and most are not, but that—simplifying—some of the processes, those belonging to the so-called *system 1* (subconscious, stereotypical and emotional), are *fast* and automatic, whereas those belonging to *system 2* (almost exactly identical with conscious problem solving),[4] will be *slow* but more *logical* (Kahneman 2011). It makes sense to relate the differences between system 1 and system 2 processes to the argument by Aristotle in his *Nicomachean Ethics*. As already stated above, he wants to hold people responsible of their actions only if they are aware of them. Now, here is the tension. The experienced car driver will normally be the better car driver just *because* most of the lower-level processes (such as operating the clutch) are automatic. He can then better concentrate (direct his conscious attention) to the road, to pedestrians, to bicycle drivers etc. This is the normal story. But let me now assume that our experienced driver makes a *mistake* in an emergency situation *because* of the automatic nature of most of the processes he employs when operating a car; say, he has to move away from a dangerous crossing, quickly pulls in the wrong gear and causes an accident because of that. A rare occasion of a failing of this type of automatic process, indeed. Is he now responsible for the accident?

I am not sure, and the problem is clearly not limited to car driving where almost everyone knows about the number of automated processes typically involved; but it easily expands to processes where an individual would not even be able to *report* on the involvement of automatic processes *ex post*. Also, whereas the automatic system 1 processes are often seen as successful, other studies have almost generally associated them with *errors* (so-called biases; Tversky and Kahneman 1974); and in those cases, the rule is *declined* decisional outcomes whereas improved outcomes are the exception.

Herdova (2016) is collecting and discussing evidence from a number of behavioral experiments, including many from social psychology (e.g., so-called

---

[4]There is no perfect overlap, however, with 'hard-problem'-type consciousness. This subtlety will be disregarded in the following discussion.

bystander experiments in which the probability of helping someone decreases with the number of other individuals that are present), demonstrating that people are largely and unconsciously driven by situational factors (i.e., some apparently peripheral characteristics of the decision situation they are in), and she is indicating that it is *debatable* whether actions that are *not* fully conscious might lead to *full* moral responsibility. Herdova (2016) ends her analysis in the following way:

> I conclude, then, that a lack of conscious awareness of the influence of situational factors, or of one's reasons for action, brought about by the situations one faces, can diminish the degree of control one exercises over one's behavior. In turn (assuming volitionism), this decrease in control mitigates one's moral responsibility—one is less responsible than one would otherwise have been. One does not, however, bear no responsibility at all for one's behavior. This is because one still exercises some degree of control over one's actions. Though lacking conscious awareness of certain things excuses us to some extent, we are still accountable for what we do. (Herdova 2016, 69)

My position, perhaps slightly leaning more towards 'excuses' than Herdova's, is that the existence and importance of unconscious decisions is *a real challenge to full responsibility*. Even with larger, more complex, high-stakes decisions where one would often assume the involvement of a large part of cognitive (system 2) processes, this might be a misspecification of the situation. In a study by Schade et al. (2012), e.g., experimental results show that even with large stakes, emotions such as worry (belonging to—at least mainly—unconscious system 1 processes) might sometimes be more important than rather cognitive processes.[5] In many of those cases, it would not be justified to hold the respective individual fully responsible for his choices as to how much emphasis to put on which reality to experience. The matter is more complex when the individual does not even know that decisions *are* high-stakes decisions or when the decision situation takes place within an everyday scenario such as car driving (see above). And although, as has been pointed out already within the discussion of the importance of *self-cultivation*, responsibility might rather work indirectly (and long-term), this clearly is not a straightforward justification for responsibility in situations where much of the (short-term) decision processes are unconscious.

Staying Away from 'Very-Low-Consciousness' Situations: Cluster Membership as a Limitation or a Basis for Responsibility?

Another potential limitation, but also a chance with respect to responsibility arises from the fact that I have assumed individuals to have a tendency to stay away from very-low-consciousness situations (see Chap. 4) or, in other words, that

[5]See, for a general overview of the decision processes involved with large stakes, Kunreuther et al. (2002).

consciousness has a tendency to cluster. What, e.g., happens, if a large consciousness cluster—given the choices of the majority—moves more and more into a direction requiring decisions from a certain individual that he finds unappealing or even unethical? Let me assume a situation where this individual has already reached the 'outer boundary' of the cluster defined by versions of individuals only having a very low level of consciousness allocated to them. Or in other words, many versions of this individual are almost 'alone' in terms of other consciousness residing in the respective realities where they reside, and even the center consciousness is operating in a scarcely populated part of the clustered-minds multiverse. Now he has to make an important choice with only two alternatives, A or B. Choosing alternative A would be just one more decision away from the cluster—think of the entire light cone moving more towards the edge of the cluster—and would lead to a situation with an uncomfortably low amount of other consciousness facing most versions (almost zombies), and even the center consciousness enters a low-consciousness situation; therefore, alternative A cannot be chosen. Alternative B would indicate a move—again, of the entire light cone—more towards the central areas of the cluster but would be, however, considered unethical by the individual. Isn't this clearly indicating a situation where responsibility is not given because the individual is, say, *forced* to move with the crowd? Or consider the opposite case: If you currently reside in a cluster characterized by highly ethical behavior, you are hardly able to do anything 'awful.' So, short-term, regarding your *next choices*, cluster membership appears to be free-will reducing and thus responsibility is reduced, too.

Long-term, however, an interesting perspective evolves, and the power of this argument is enhanced by the possibility, introduced in Box 6.3, of implementing ex post reallocations of consciousness[6]: Self-cultivation à la Confucianism or self-forming actions à la Kane might *bring you into a certain minds cluster*. Rather than using, e.g., cluster membership as an excuse for bad actions, individuals might then be viewed as being (partially, at least) *responsible for the minds cluster they reside in* (limiting their possibility of performing bad and good actions). The latter perspective offers a new basis for responsibility: Residence in a certain minds cluster becomes a self-binding commitment à la Schelling (1981 [1960], 1978). Entering a certain minds cluster and maintaining membership in that cluster is probably a matter of carefully targeted, multiple subsequent choices, made over a larger period of (subjective) time, plus multiple reallocations of consciousness. It is hard to understand how this exactly works because we are neither aware of reality clusters, nor do we consciously realize what it means to stay clear of very-low-consciousness situations.

As an intermediate summary, staying away from very-low-consciousness situations, together with, perhaps, 'meaningful clustering' (for more on this term, see Chap. 10) and possible reallocations of consciousness, might have a two-fold effect

[6]Indeed, unexpected changes in the development of a minds cluster are probably one of the major reasons for reallocations of consciousness.

on responsibility. Short-term, cluster membership potentially limits the freedom I have to make 'good decisions,' say, in a 'bad cluster,' and 'bad decisions' in a good cluster—and hence my responsibility for those choices. Long-term, responsibility might partially *pertain* to cluster membership. Self-cultivation, self-forming actions and adjustments made via reallocations of consciousness are about long-term orientation, about multiple decisions, and, thus, about which minds cluster to reside in. This is close to how responsibility is dealt with in the eastern traditions and quite far away from the western law system, I admit. So how should we 'operate,' here? Box 7.1 deals with a quite unusual approach in this regard.

Box 7.1: 'Instead of responsibility'—The economics-based approach by Becker (1974)

Gary S. Becker, Nobel laureate in economics, generally proposes behavioral modeling according to economic theory and economic incentives for the sake of a better understanding of and for developing means to change individuals' actions. With respect to crime and punishment, his question is *not* about *responsibility*, but rather about the *costs and benefits of committing a crime*. This might be more helpful an approach the more ambiguous the assessment of responsibility in certain decision situations is. Within a society-level analysis, Becker proposes the following calculus of an individual that considers committing a 'crime:' "The approach taken here follows the economists' usual analysis of choice and assumes that a person commits an offense if the expected utility to him exceeds the utility he could get by using his time and other resources at other activities [see Chap. 8, for more details on the expected utility framework]. Some persons become 'criminals,' therefore, not because their basic motivation differs from that of other persons, but because their benefits and costs differ. (…) This approach implies that there is a function relating the number of offenses by any person to his probability of conviction, to his punishment if convicted, and to other variables (…)" (Becker 1974, 9; insertion by the current author). This approach might be considered 'cold,' on the one hand, and I have to admit that I agree with this critical view, emotionally. On the other hand, it offers a responsibility-independent justification of the law system and of punishing certain criminal activities. People are not punished because there are held responsible for a 'crime,' they are punished because this is part of an incentive system reducing the general level of criminal activities.

General Conclusions

What is the general summary that can be offered and what are the conclusions of the somewhat complex analysis presented in this chapter? I started out with a couple of different philosophical perspectives on responsibility. Some of those theories, but also specific thoughts within those theories, were closely related to the discussion on free will, others turned out to be fully independent of the discussion in Chap. 6. Some theories were dealt with in more detail. Two of those were selected from different periods of western thought (Aristotle as well as Strawson), two were selected from eastern religious (or spiritual) traditions (Buddhism as well as Confucianism). Relating those theories to the situation described by the clustered-minds multiverse, some difficulties with single-decision (or short-term) responsibility were detected because of the only indirect coupling of preferences with what different versions of the individual will do—likely with low or medium-level consciousness. At this point it is important to recall that extreme allocations of consciousness cannot simply be required from the decision maker (according to premise 5 of the clustered-minds multiverse; see Chap. 4). On the other hand, long-term developments of a person could more easily be related to responsibility because of multiple, repeated choices and reallocations of consciousness being the basis of individuals' residence in different minds clusters—having an impact on their singular choices in turn. Cluster membership might be described as a self-binding commitment à la Schelling (1981) [1960]. One more potential caveat to responsibility, however, was detected and discussed in this chapter, unconsciously driven choices caused by, e.g., situational factors.

Responsibility in the clustered-minds multiverse turns out to be a difficult, but clearly not a hopeless subject. It is possible to hold individuals (at least partially) responsible for the development of their lives and their 'general orientation,' being in close correspondence to cluster membership, say, their *propensity* to engage in criminal action—and for some individual decisions resulting from that; but it is also difficult to hold them responsible for other individual decisions, made in proximity to the edge of consciousness clusters, and also for partially unconscious ones, resulting from situational factors.[7] Since compatibilist arguments, crafted for the situation in a singular, deterministic universe, have been shown not to be convincing, at least not to the current author, the situation for responsibility is to be judged clearly more favorable in the clustered-minds multiverse than in such a deterministic, singular universe.

But nevertheless: Individuals cannot be held fully responsible for each and every decision they have made (this would also apply to most compatibilist accounts of responsibility; for good reasons, Frankfurt's as well as Strawson's approaches are 'case-based'). This might not be a problem for life in general, where, again, the situation for responsibility is much improved over the situation in a singular,

[7] And I have to admit that it might often be impossible to judge which are the respective conditions at hand, i.e., which of those situational determinants apply.

deterministic universe, but perhaps for jurisprudence, because the determinants of responsibility that I have characterized, based on the clustered-minds multiverse, are very *different* from those within the Frankfurt cases and Strawson's analysis and far less easy to determine—to say the least. What could be helpful, here? Could an economic account such as that sketched in Box 7.1 be a way out? Further research and intensive discussions are necessary, here.

PART III
QUANTUM DECISION MAKING IN THE MULTIVERSE AS VECTORIAL CHOICE: TOWARDS A TRANSFORMATION OF THE DECISION SCIENCES

Chapter 8
General Framework, Objective Function, and Probability

Normative Decision Theory

This chapter is the first of three chapters that explore consequences of the clustered-minds multiverse for the decision sciences. The impact should be expected to be substantial, given the fact that the debate on consciousness and free will was strongly influenced by this new theoretical standpoint in the last chapters and the decision sciences are concerned with a closely related subject: *choices* made by individuals—choices in the 'standard' meaning within the decision sciences and in the meaning, defined in Chap. 4 and (partially, for the case of free will) in Chap. 6, in the clustered-minds multiverse. The current chapter concerns itself with the general framework of the decision sciences, with the objective function that is typically used and its appropriateness, using a multiverse perspective, and with macro-world probabilities in the sense of a lack of knowledge of future conditions. The next chapter then analyzes different existing concepts of utility as well as potential modifications and how they match with the conditions within the clustered-minds multiverse. The subsequent chapter finally looks at strategic decision making and economic models and how the perspective on them changes. All three chapters take the normative, i.e., prescriptive, as well as the behavioral, i.e., descriptive sides of the decision sciences into account (see, for more details with respect to this classification, Kleindorfer et al. 1993).

The beginning of the modern decision sciences might be identified with the work by Bernoulli (1954 [1738]) in his attempt to address the so-called *St. Petersburg paradox*. In the St. Petersburg game, a fair coin is tossed. One side of the coin, side *A*, is associated with doubling the money in the pot starting with, say, five dollars. So, if side *A* of the coin comes up once, the amount is doubled to ten dollars, if it comes up another time, it is doubled again to twenty dollars. If it comes up three times in a row, the amount in the pot will be forty dollars. And this doubling of the money *may* continue an infinite number of times *if* side *A* of the coin comes up uninterruptedly for an infinite number of times. However, the series stops whenever

© Springer Nature Switzerland AG 2018

C. D. Schade, *Free Will and Consciousness in the Multiverse*,
https://doi.org/10.1007/978-3-030-03583-9_8

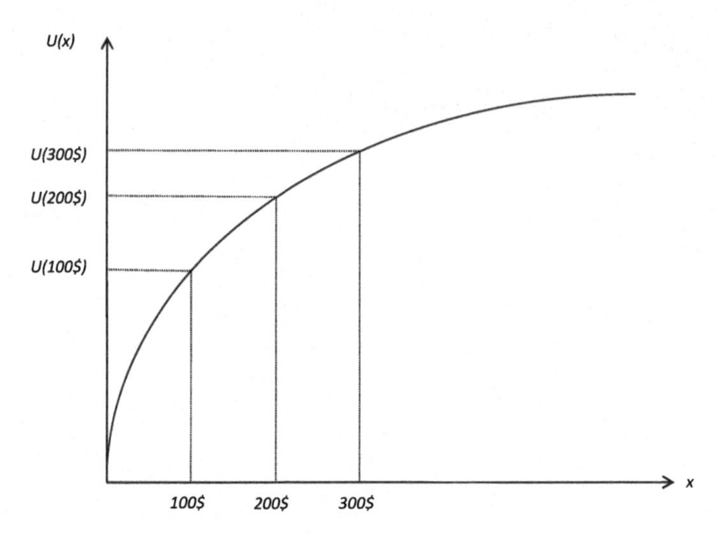

Fig. 8.1 A concave utility function

side *B* of the coin comes up. When side *B* comes up right away, you just get the starting amount in the pot: five dollars. When it comes up after side *A* came up twice, you get the twenty dollars in the pot etc. Mathematically, the expected value of the game can be shown to be *infinite*; but there is a large chance that only *small amounts* will be paid out. Consequently, people are willing to pay only small amounts to play that game. The situation appears paradoxical if *rational decision making* is associated with maximizing *statistical expected value* (what has been the theoretical benchmark at Bernoulli's time). Given that the expected value of the St. Petersburg game is infinite, how come that individuals are not even close to betting their entire wealth on it?

The way that Bernoulli proposed to address the paradox was to introduce a so-called *utility function* that is *concave* for a typical (i.e., risk averse) decision maker.[1] A concave utility function is shown in Fig. 8.1. With such a function, the increase in utility $u(200\$)$–$u(100\$)$ from an increase in wealth from 100$ to 200$ is larger than the increase in utility $u(300\$)$–$u(200\$)$ resulting from an increase from 200$ to 300$ or even from 100,000$ to 100,100$ (this last example not shown in Fig. 8.1), although the difference is always 100$. Applying this principle to the St. Petersburg game makes the paradox disappear because it is now *rational* to only pay small amounts for the participation in that game. Specifically, if not the potential monetary outcomes x of the game are weighted with their respective probabilities p of occurrence directly (e.g., with $p = .50 \cdot .50 = .25$ for getting 20$),

[1]Empirically, about 80% of individuals are risk averse and exhibit a concave utility function. About 20% are either risk neutral or risk taking exhibiting either a linear or a convex utility function. For an introduction into the relevant literature see Holt and Laury (2002), Harrison and Rutström (2015).

Table 8.1 Basic ingredients of normative (singular-attribute) decision theory under risk

Goal	Basis	Definition of welfare/ outcomes	Modelling of utility	Modelling of probability	Evaluation of one alternative
Maximization of an individual's welfare	Axioms of rationality	Utility of overall wealth level	Monotone increasing, convex or concave utility functions	Linear processing of objective or subjective probability	Sum of probability-weighted utilities of potential outcomes

but instead the utilities $u(x)$ of those outcomes, it can be explained why many people would *not* put their entire wealth at stake as would be predicted based on the statistical expected value. The replacement of expected value with *expected utility* accounts for risk aversion.

Bernoulli's idea has been refined in the middle of the last century by showing that expected utility theory can be mathematically derived from a small set of plausible *rationality axioms* such as those postulated by von Neumann and Morgenstern (1953) [1947] or those underlying *subjective expected utility* theory by Savage (1954), the latter allowing for 'non-technical' probabilities (such as those in real-life choices, i.e., not employing fair coins, roulette tables etc.) and subjective probability estimates by the decision maker.

Several alternative axiom systems have been proposed, some with more, some with fewer axioms. But all those approaches, forming the class of *standard normative decision theory*, end up with the following basic ingredients: a monotone increasing utility function, either convex or concave (the theoretically important but empirically rare case of risk neutrality implies a linear function), a direct multiplication of outcomes with objective or subjective (non-weighted) probability, mostly interpreted as a lack of knowledge about future developments, and the summation of all those terms (utility in a specific state of nature multiplied with the probability of that state of nature; those terms then summed over all states of nature). A singular alternative, the one with the *highest expected utility* (ordering the results of the summation) is supposed to be selected, and since this principle is based on rationality axioms, obeying to its implications is seen as inevitable for a rational decision maker.[2] That is the reason why the theory is called *normative*: a rational decision maker *should* obey to it (at least if he agrees with those rationality axioms—what most decision makers do).[3] The basic ingredients are systematically displayed in Table 8.1.

[2]A great introduction into normative and descriptive decision theory is to be found in Kleindorfer et al. (1993).

[3]Modifications of the normative benchmark applying weaker rationality assumptions within generalized utility models (e.g., Machina 1982, for the case of the independence axiom) will not be dealt with in this book.

Note that in standard (and subjective) expected utility theory, a *singular utility dimension* is assumed to exist (or to be constructed); whereas there are also several multidimensional approaches existing (see, for an overview of the large class of multicriteria-decision-making models, e.g., Triantaphyllou 2000; for adequacy-importance modeling or the theory of reasoned action Fischbein and Aizen 1975; for multidimensional utility in consumer demand Lancaster 1966). (The multidimensional character pertains to one reality, I should add, because a different type of multidimensional utility will be introduced for the multiverse, below.) In many cases, however, those models do not incorporate risk (or the modelling of risk is somewhat unsatisfactory). In this chapter, I will not be looking at multi-attribute models. They will be dealt with in the next chapter.

Box 8.1: Quantum-world and macro-world probabilities

The translation of quantum-world into macro-world probabilities is *practically impossible* outside the experimental physics laboratory (or apart from a close coupling of macro-events to events in that laboratory; see, e.g., Schrödinger's cat thought experiment, described in Chap. 2). Comparisons between those two types of probabilities are also *logically problematic*, since quantum-world and macro-world probabilities have a different ontological and epistemological status. Whereas quantum probabilities result from an auxiliary equation to the Schrödinger equation, the Born (1926) rule (and are hence part of quantum calculus), generating relative frequencies of measurement outcomes along one decoherent history (see Chap. 4), macro-world probabilities ('decoupled' from quantum events in laboratories) are a way of capturing individuals' ignorance with respect to future developments in the macro world. This has to be kept in mind when dealing with probabilities within (normative) decision models, looking at them from the perspective of quantum mechanics.

The Schrödinger equation is deterministic, but this does not mean that individuals have a clear picture of the future. And according to Box 8.1, we cannot normally use information from the Born rule to inform our macro-world predictions. The frameworks of standard expected utility theory and subjective expected utility theory *have* been developed for classical probabilities of future developments in the *macro sphere*. In this regard, normative decision theory might be seen as an appropriate framework, also for 'most' macro-conditions that pertain to the clustered-minds multiverse; as has also been pointed out in Box 8.1, probabilities in the macro sphere can, for the majority of cases, only be interpreted in the form of a *lack of knowledge*.[4] They are a 'tool' that helps us tackling the uncertainty we face when making a choice due to this lack of knowledge. Hence, there is not much of a

[4]The 'majority of cases' exclude the rare, artificially constructed cases where a direct coupling of macro to quantum events (see, again, Box 8.1) is possible.

difference, whatsoever, between a singular universe and the quantum multiverse with respect to the usage of macro-world probabilities. There are other reasons as to why the framework of normative decision theory might not be appropriate for the situation in the clustered-minds multiverse, as will be demonstrated later in this (and the next) chapter, however.

The chapter proceeds as follows. The next section deals with behavioral decision theory, taking into account real decision makers' deviations from rationality and approaches of quantum decision making suggested by other authors. The subsequent section deals with a different framework of decision making: effectuation. The core section is the subsequent one. Here, substantial changes in the general setup of decision models are proposed, based on the clustered-minds multiverse. Far from offering a conclusive account, the next section then deals with possible advantages or disadvantages of effectuation—as compared with other, more traditional, decision-making frameworks, when looking at all of the approaches from the perspective of the clustered-minds multiverse. The chapter ends with a final discussion and conclusions section.

Frameworks of Behavioral Decision Theory and Quantum Decision Making in the Form Proposed by Other Authors

Normative (subjective) expected utility has originally been formulated independently of any physical basis. Recently, however, the Oxford school (see, e.g., Wallace 2012c) has proposed a decision-theoretic foundation of the Born (1926) rule, that has been criticized for conceptual reasons in Chap. 4. Although linking quantum mechanics and the decision sciences, this approach is unrelated to what will be proposed later in this (and the next) chapter(s).[5]

Normative expected utility theory has been *criticized* in its singular-universe usage as an account of individuals' *actual behavior*. (Some proponents of this theory would claim that *prediction* is, anyway, not the purpose of this theory, but rather *prescription*.) The criticism is about as old as its several, closely related axiom systems are. According to Allais (1953), it can be shown that most people clearly violate at least one of the rationality axioms of (subjective) *expected utility theory*, the so-called *independence axiom* (appearing, with similar labelling, in most axiom systems), that they change their preferences in an irrational way if the *same outcome* is added to two lotteries.[6] Let me suppose than an individual has been asked to rank-order, according to his preference, two lotteries A and B and that

[5]The Oxford-School approach simply has a different purpose and does not allow for insights into actual decision making in the multiverse.

[6]The simplest case of a lottery is two different outcomes occurring with non-degenerate probabilities p and $1 - p$; whereby a non-degenerate probability means that nothing occurs with certainty ($p = 1$) or not at all ($p = 0$).

he reports to prefer, say, A over B. Now an identical outcome is added to both lotteries and he is supposed to rank-order the two lotteries again. And now, under certain experimental conditions (that are a bit too complex for a quick exposition, here), he reports a preference toward lottery B' (the transformed lottery B) over A' (the transformed lottery A)—but he *should* not because this is irrational according to expected utility theory.

Another important violation of rationality has then been shown by Ellsberg (1961) that pertains to *subjective expected utility theory*. It could be demonstrated, empirically, that most individuals are *not* indifferent between lotteries containing exact probabilities and lotteries containing 'uncertain probabilities' of outcomes. Indeed, most individuals systematically dislike probabilities that are not exactly known. The phenomenon has been labelled *ambiguity aversion*. Note that according to subjective expected utility, a rational decision maker should not make a difference between situations of ambiguity and exact probabilities as long as the average probability estimates in the ambiguity situation equal the exact probabilities (Ellsberg 1961).

A huge stream of literature has evolved out of those initial papers. The respective field is called *behavioral decision theory* (or behavioral economics or economic psychology, depending on some nuances and the authors as well as their affiliations). The most influential works so far have been those by Daniel Kahneman, Amos Tversky and Richard Thaler (e.g., Kahneman and Tversky 1979; Tversky and Kahneman 1974, 1992; Thaler and Johnson 1990). Summarizing the most important findings of those works, individuals have been shown to apply *heuristics* (simplifying: 'rules of thumb') to complex decision problems and to fall prey to certain systematic *biases*—leading to deviations from rational behavior (Tversky and Kahneman 1974).

Furthermore, when analyzing problems within a framework similar to that suggested within expected utility theory but taking into account real individuals' behavior, a model called *prospect theory* (including cumulative prospect theory) results (Kahneman and Tversky 1979; Tversky and Kahneman 1992). According to prospect theory in any of its versions, individuals do not care about the utility of their *total wealth* but their decisions are rather driven by *changes* (gains and losses) relative to some *reference point*. The reference point might in turn be determined by either the wealth in the current situation (but subjectively set to zero since it is the status quo), or by some (short-term) goal that has or has not been reached so far. The so-called *value function* is then convex below, but concave above the reference point, indicating *risk aversion with gains*, however *risk taking with losses*. Finally, the function is *steeper with losses* than gains (so-called *loss aversion*).

Probabilities are not processed the way they are (as would be the case in expected and subjective expected utility theory), but they enter a *risk-weighting function*. This function differs between the standard and the cumulative version of prospect theory, but it is never linear. In the standard version (Kahneman and Tversky 1979), e.g., small probabilities are either overweighted or treated as being zero (leading to a discontinuity in the function), medium-level probabilities are rather underweighted etc. The value function 'replaces' the utility function in

Table 8.2 Basic ingredients of singular-attribute prospect theory, the most influential behavioral decision theory under risk (regular version: Kahneman and Tversky 1979; cumulative version: Tversky and Kahneman 1992)

Goal	Basis	Definition of welfare/ outcomes	Modelling of utility	Modelling of probability	Evaluation of one alternative
Understanding and predicting how individuals make choices	Experimental findings	Outcomes are defined (as gains/ losses) relative to a reference point	Value function: convex below and concave above the reference point; steeper in the loss domain	Non-linear weighting function: processing of objective or subjective probability	Sum of the products of the results of the value function and the results of the probability weighting function

expected utility theory, the risk weighting function replaces the probabilities. The basic ingredients of (singular-attribute) prospect theory[7] are depicted in Table 8.2.

Whereas many researchers involved in the decision sciences believe that prospect theory, especially in its novel version of cumulative prospect theory (Tversky and Kahneman 1992), offers a satisfactory account of important behavioral phenomena in decision making under risk, others have debated this and developed alternative models.[8] Two closely related strands of literature are often referred to as 'quantum decision making' (e.g., Busemeyer et al. 2009) and 'quantum social science' (e.g., Haven and Khrennikov 2013). But they have a very different take on the link between quantum mechanics and decision making as that proposed in the current monograph. Haven and Khrennikov (2013) are especially explicit about just applying the *quantum formalism* to human decision making whilst avoiding to answer the question why it should be relevant to it:

> We emphasize that in our approach, the quantum-like behavior of human beings is not a consequence of quantum-physical processes in the brain. Our premise is that information processing by complex social systems can be described by the mathematical apparatus of quantum mechanics. (Haven and Khrennikov 2013, xviii)[9]

[7]Multi-attribute versions of prospect theory will not be dealt with in this book.

[8]An overview and discussion of the different types of alternative theories and models that have been proposed in the literature are beyond the scope of this book. But it should be mentioned, here, that an approach somewhat opposed to Kahneman and Tversky's type of reasoning has been quite successful, too, the approach by Gigerenzer and coauthors (see, e.g., Gigerenzer and Todd 1999). According to my reading, the recent book by Kahneman (2011) takes a middle position between his and Tversky's older and Gigerenzer's approaches. The last section of this chapter briefly touches upon Gigerenzer et al's approach.

[9]I should also note that modelling both within quantum decision making and quantum social science is based on the Copenhagen-type and Born (1926) rule formalism and not on the multiverse interpretation of quantum mechanics.

This book is not looking at quantum-mechanical processes in the brain. It is also not trying to implement, in the form of a purely 'technical' usage, a quantum-mechanical calculus for descriptive purposes. So, although the approach to decision making to be developed based on the clustered-minds multiverse in this and the following two chapters is called *quantum decision making* and although some of the thoughts to be developed in Chaps. 11 and 12 might be called *quantum social science*, not much can be learned, conceptually, from the existing approaches using those names for the approach pursued in this book.

Causation Versus Effectuation

Based on Simon's (1955, 1957) idea of *bounded rationality*, Sarasvathy (2001) has proposed a principle of decision making: *effectuation*, that is supposed to complement (or perhaps replace) the classical normative approach of decision making; in fact, it also *criticizes* the normative approach of decision making as being inappropriate for many realistic decision situations; and it also (implicitly) rejects the feasibility—again, for many realistic cases—of behavioral approaches of decision making that keep the *structure* of normative models such as prospect theory. In a way, Sarasvathy's approach might neither be seen as entirely normative nor as entirely descriptive. It rather *describes* the actual decision processes employed by successful decision makers (for the case of experienced versus novice entrepreneurs see Dew et al. 2009) and *suggests* to others to use the same decision processes as well to be successful. Although crafted for the analysis of entrepreneurial behavior, it is applicable to human decision making more generally.

There are systematic differences between the so-called *causation* principle, the term used by Sarasvathy and coauthors for the type of decision making explicated in its 'pure' form in the first section of this chapter and for other, normative decision principles based on maximization including multidimensional approaches, and the so-called *effectuation* principle (e.g., Sarasvathy 2001; Dew et al. 2009; Faschingbauer 2013). As depicted in Table 8.3, causation and effectuation indeed vary dramatically on several dimensions; note that the dimensions used to describe the differences between the two approaches differ from those used to describe normative and behavioral decision theories in Tables 8.1 and 8.2.[10]

The differences between the two principles reported in Table 8.3 will be described using two related examples. Note that the two examples are necessarily only related and not identical since the application of the two decision principles typically lead to different *aspects* to be decided upon (Sarasvathy 2001); the entire decision situation exhibits a causation or effectuation *structure*: (1) A typical example for the use of the causation principle from the field of economics would be

[10]The classifications/comparisons would become far more complex if the same set of criteria were to be used for both purposes.

Table 8.3 Causation versus effectuation principles (based on Sarasvathy 2001; Faschingbauer 2013, 30–34; the latter especially for the future/probabilities column)

Criteria/ principles	Starting conditions	Constraints	To be decided upon	Future/probabilities	Objectives	Selection criterion
Causation	Given goal	On possible means	Set of alternative means	Future predictable probabilistically/only the predictable can be steered	Expected return	Selection between means via maximization
Effectuation	Given set of (unalterable) means	On possible effects	Set of possible effects	Future unpredictable/everything that can be steered does not have to be predicted (even probabilistically)	Keep investment/ loss affordable; 'survival'	Selection between effects via satisficing

entry with some radical technological innovation into a novel market by some entrepreneur after market research, following a textbook approach. A goal would be pursued to start with and that goal might be maximizing the expected return (a profit measure) from market entry with the *pre-specified* technological innovation. The constraints might pertain to the available entry strategies (given, say, the knowledge base of the entrepreneur and his employees), and one of those strategies has to be decided in favor of (perhaps including the alternative of no entry). Potential future developments, their consequences and their probabilities of occurrence, given the choice of an entry strategy, are supposed to be estimated as precisely as possible. That is the reason why (extensive and expensive) market research has been conducted to start with. All that has to be done for all available entry strategies; the criterion of selection between them is *maximization*. The entry alternative that maximizes expected return has to be chosen. (2) Now let me look at the workings of effectuation for a related decision.[11] As typical for effectuation, the exact goal is not pre-specified, so I am limited here to saying that I am looking at 'innovative behavior' of an entrepreneur. The starting conditions are now the resources and knowledge this person possesses. The constraints are not the available strategies (as before), but quite similarly, the cost and feasibility of different entry strategies given the fact that the entrepreneur might not be able to generate enough credit to finance some of them or that he does not know enough to pursue them even if the monetary issues were addressed. Moreover, *there is no given innovation that has to be brought to market*; instead, the product we eventually sell is open. Indeed, that is what we really decide upon: *What* to market (instead of how to market, in the case of causation). So, given our restrictions, we might come up with all kinds of possible outcomes in the *feasible range*.[12]

Whereas the causation principle aims at estimating the probabilities of occurrence of certain events as precisely as possible (the ideal being a 'technical' probability), supposing that this is possible and that this is what we *can* do to *control* the future (and nothing else), the effectuation principle would imply that such an estimation is often difficult to impossible in non-trivial (i.e., typical for real-life) situations of uncertainty. One would, using this approach, rather try to *steer* the future in other ways as much as possible. Oftentimes, the causation

[11]It has already been mentioned that applying causation and effectuation may lead to a different decisional structure and to different decisional outcomes. However, some authors contributing to the effectuation literature would possibly argue that causation and effectuation are typically applied to different decision situations to start with. That might sometimes be correct. I am not so sure, however, whether it always makes sense to 'allocate' the two principles to different situations. I am also not sure whether there is such thing in practical decisions as a sole use of one or the other principle. Frankly, there might not be any decision without the (partial) application of effectuation.

[12]Another, somewhat casual and non-economic example for column four of Table 8.3, comparing the principles of effectuation and causation, is cooking a dinner for friends (see for this as well as the related example of "curry-in-a-hurry," Sarasvathy 2001). A decision maker applying causation picks a meal from the cookbook, buys the ingredients at a grocer and prepares them according to the recipe. A person applying effectuation looks at what is in the fridge and in the drawers, tries to combine the possible ingredients to creative tasty dishes and cooks one of them.

principle would advise to aim at *protecting* against unpredicted developments, whereas the effectuation principle would rather prescribe to try to find other ways to deal with them (Faschingbauer 2013, 66–68). The objective of effectuation is to *limit* the possibility of negative developments, to keep the investment affordable and to keep the potential loss affordable (two related things). If we manage to keep those two in an affordable range then we *satisfice* (there is no maximization involved). *Not the ex-ante maximization of expected outcomes but the ex-post survival is at the core.*

There is another way of describing the advantage of the effectuation approach. It is a smart way of dealing with the *real-options structure* of many real-life decisions (Dixit 1992; Dixit and Pindyck 1994). This thought will further be explored in the next chapter, because different utility concepts *appear* to evoke this issue to a different extent. Here, it may suffice to make mention of the fact that real-options reasoning takes into account the effect of *irreversible actions* (or commitments) on present choices, and that an explicit modelling of such a structure is already complex in a singular universe. With respect to a realistic modelling of choices in the multiverse, such a structure might often be impossible to implement. But there is also good news pertaining to the multiverse framework. The potential problem with real-options structures is generally alleviated by the possibility of ex-post reallocations of consciousness (see Box 6.3).[13] In any case, even without a solution to the modelling problem, effectuation offers a good *heuristic* for dealing with (potential) real-options structures, to engage in only stepwise, smaller commitments rather than producing high-stakes situations with early points of no return.

In line with those considerations, there are some further consequences for the example that I have used: Whereas the causation principle might lead to the selection of an expensive market entry, huge credit to develop and market a radical innovation with large chances of big success and big failure (i.e., situations of substantial, early irreversibility), the effectuation principle might lead to a less radical, rather stepwise innovation process where partnerships are looked for, (large) credits avoided, and each and every step tested with the actual market in a trial-and-error approach (Sarasvathy 2001). This does not necessarily imply that the *final* innovation will be less radical, only the stepwise process will confront the decision maker with a situation where *each single step is less radical.*

It might appear as if the evaluation of causation and effectuation approaches is somewhat 'balanced,' as if sometimes causation, sometimes effectuation is more appropriate a strategy under uncertainty (in the sense of successful decision making, i.e., as a good prescriptive approach) or whether the results of the two are so different that it is eventually a matter of taste which one to use. This is, however, *not* the way Sarasvathy (2001) and Faschingbauer (2013) see it in a singular universe. They somewhat 'prefer' the effectuation approach, at least for a large set

---

[13]The problem is alleviated, not solved, given the fact that we do not know how smooth and thorough reallocations of consciousness are.

of cases they have in mind.[14] From the perspective of the clustered-minds multiverse, the judgment has to be postponed until the section following the next.

Changing the Objective Function: Allocation of Consciousness to Different Alternatives Instead of Choosing *the* Utility Maximizing Alternative

There is something fundamentally different in the clustered-minds multiverse rather than in a singular universe that cannot be accommodated for within any of the decision-theoretic frameworks that have been presented and discussed, so far. The problem pertains to the objective function as well as to the way one supposes decisions to be implemented. As has already been pointed out within the previous chapters, individuals are *not deciding*, in a narrow sense, in the form of selecting and implementing the 'best' alternative; and this causes trouble for the application of existing decision models, independent of whether optimization is assumed in the normative variants, satisficing in effectuation etc. Instead, phenomenologically, hidden to the singular-world consciousness of a specific version of an individual currently experiencing one reality, consciousness of the individual is allocated to different versions to a different extent; and each version of the individual experiences himself making just one choice. How would current decision theory have to be changed to accommodate for this?

Implementing the novel perspective on life, suggested within the clustered-minds multiverse, leads to a number of premises with respect to the decision sciences, to be outlined in the following. Note that I will introduce and use, from here on, the terms *vectorial decision* or *vectorial choice* (synonymously) to indicate that different weights are put on different realities rather than a singular reality selected. In other words, the term vectorial decisions (in contrast to, say, 'regular' decisions in a singular universe) is supposed to express that there are multiple realities where consciousness will be allocated to with different weight. So, the outcome of a decision is a vector, the components represent the intensity of consciousness in different realities.

The idea of a vectorial choice is formally depicted, in a simple form, in the following. In Formula (8.1), the vector stands for the outcome of some vectorial decision (*VD*), the allocation of consciousness to different realities. Each component a_i describes the amount of consciousness allocated to a specific reality or the *emphasis/awareness* this reality receives. The number of components is arbitrarily large and equals the number of realities concerned (n), but the weights put on different realities must add up to one; this assumes that consciousness can only be

[14]Pushing it to the extreme, the market introduction of a novel, high-development-cost pharmaceutical, relevant for the global market, is probably a case where Sarasvathy and Faschingbauer would advise the usage of causation.

distributed between different realities, and that there is a total amount of consciousness that cannot be surpassed (8.1). Note that this total amount of consciousness is always the (sub-) amount of the entire consciousness of an individual that pertains to that version of the individual that becomes the decision-making 'individual' in the respective next decision to distribute consciousness to different realities. (The complexity arises from the fact that this is a branching structure with individuals splitting into versions and those versions playing the role of the 'individual' in the respective next decisions.) Moreover, as has also been analyzed in Box 6.3, this again demonstrates that consciousness (on the level of *all* versions of the individual) has to regularly be reallocated not to run into cases of extreme dilutions of consciousness.[15] In the example, described in Formula (8.2), of a vectorial decision of, say, Louise (VD_L) there are three different realities (buying a Tesla in a_1, a Mercedes in a_2 and a BMW in a_3), and most weight is put on the second reality (specifically, 70% of overall consciousness is allocated to the second reality where Louise buys the Mercedes). In line with Formula (8.1), the allocations to the three different realities add up to 100%.

$$VD = \begin{pmatrix} a_1 \\ a_2 \\ \dots \\ a_i \\ \dots \\ a_{n-1} \\ a_n \end{pmatrix} \in R^n, \quad \text{with } a_i > 0 \text{ and } \sum_{i=1}^{n} a_i = 1 \qquad (8.1)$$

$$VD_L = \begin{pmatrix} a_1 \\ a_2 \\ a_3 \end{pmatrix} = \begin{pmatrix} 0.1 \\ 0.7 \\ 0.2 \end{pmatrix} \qquad (8.2)$$

And these are the premises, resulting from this new view:

(1) Our occupation with decision models is an *aid* to allocate the appropriate amounts of consciousness to different realities. Decision models consult consciousness in making vectorial decisions. However, decisions are never *directly* implemented.
(2) Decision scientists as well as decision makers have to accept the fact that the selection of an alternative and the later implementation of this singular choice per se is an illusion—it only pertains to the version of the individual that is currently experienced.
(3) Since the experienced version's maximization of expected utility is, anyway, an illusion, because not a singular reality is selected by one version of the individual (or just by 'the' individual, in a singular-reality view) but a vectorial choice made by the individual facing multiple

[15]Since those reallocations may or may not take place continuously (or rather from time to time) it is unclear whether overruns of the total amount of consciousness might temporarily be admitted.

realities, one might want to replace or at least supplement it with a more open simulation of the effects that different 'choices' might have to best inform the appropriate *allocation* of consciousness instead. This is especially relevant as long as potential *utility interdependencies* between different versions of the individual are not well understood or even appropriately modelled and an overall utility for the individual (across different versions) thus cannot be determined. This is also relevant because the *exact restrictions* that pertain to the allocation of consciousness are unknown and potentially complex. Known restrictions, from the premises of the clustered-minds multiverse (albeit not in the precision of a formula), are the avoidance of very-low-consciousness situations and the tendency to look for moderate allocations of consciousness (the latter perhaps in an interplay with other cluster members).

(4) Models with an open structure such as effectuation as well as multi-criteria decision models (see the next chapter) might be the 'lower-hanging fruit' with respect to the development of a quantum decision making framework and more helpful, at least short-term, for application purposes than mathematical, singular-utility optimization models (such as expected utility theory).

(5) One might expect further, currently unforeseeable effects, related to the utility interdependencies between different versions of the individual just mentioned. E.g., if the conflict between different aspects of several realities (or decision alternatives, for that matter) is large, consciousness might be 'more split' after a decision between different realities (in the sense of being less concentrated with a few) than if there are hardly any conflicts. It might not matter much in this regard whether or not we are able to mathematically account for, say, risk-reward conflicts (generating one optimal alternative) or able to weigh different utility dimensions of alternatives differently. Those models might be too restrictive—at least in their current version—to account for the complexity of vectorial choices.

(6) Although highly opaque, better understanding (or at least smartly speculating about) the distribution of consciousness based on the utility of different courses of reality experienced, might be helpful in providing decision aids, based on the different purposes that consciousness of an individual might pursue. E.g., if the aim of the individual, for whatever reason, is to concentrate consciousness with certain realities, however restricted by the requirement to avoid extreme allocations (see point 5 in the list of premises of the clustered-minds multiverse in Chap. 4), it might be more helpful to resolve (or circumvent) conflicts in decisions rather than calculating an optimum that will, anyway, not be implemented with a high concentration of consciousness.

(7) A way to *circumvent* conflicts, e.g., between risk and return of different alternatives, might be effectuation where harsh tradeoffs are typically avoided via a stepwise approach to decision making ('testing the waters;' see the next section). Another way of circumventing conflicts between different advantages or disadvantages of alternatives might be looking for compromise alternatives instead of optimizing, say, within an adequacy-importance model (for that type of model see the next chapter).

(8) Somewhat speculative, I suppose, there might be several factors impacting on the actual allocation of consciousness to different realities (i.e., on vectorial choices) that are only partially under conscious control of the individual making use of decision models—the *final* implementation of the distribution of consciousness is, anyway, never fully conscious since we only experience one reality, before and after that choice. An overview of a *subset* of such potential factors is provided in Table 8.4. Table 8.4 follows a descriptive and a prescriptive approach—it finally tries to be helpful for the decision maker, albeit currently rather in the stage of development. Note that, in order to reduce complexity, the requirement of implementing rather moderate allocations of consciousness (Chap. 4, premise 5 of the multiverse) has been left out of the considerations within that table. Moreover, reallocations of consciousness (according to Box 6.3) are not *explicitly* dealt with. But in a way, the criteria listed in Table 8.4 (with only few modifications) might also be helpful with understanding 'internal' reallocations of consciousness, decoupled from concrete decisions. (Note that the latter would imply that reallocations of consciousness would as much be driven by unconscious motives etc. as the regular vectorial choices; or in other words, free will would not account for the major part of those reallocation decisions.)

Table 8.4 allows for an insight into the complexity of *vectorial decisions* in the clustered-minds multiverse as well as the potentially small—depending on the relative importance of all the factors listed in Table 8.4—impact of free will in some decision situations. Note that there are still several factors missing. One was already mentioned above: the requirement of moderate allocations of consciousness. Others are, e.g., intuition, emotion and decision heuristics (the latter mentioned in more detail above in connection with the behavioral decision models). According to the reduced set of factors listed in Table 8.4, the consciousness distribution vector (vectorial choice) is influenced by the following factors: unconscious motives of the decision maker; unknown motives of other decision makers; impact of other, including non-human factors; utility interdependencies between different versions of the individual; decisional conflicts; and free will (as briefly as well as only partially noted in Table 8.4, consciousness prepares the 'free-will' choice, but the final implementation of the vectorial choice is unknown to the decision maker).

Regarding the unconscious motives, Chaps. 9 and 11 will provide some applications of those types of influences. E.g., repetition compulsion, a concept from

Table 8.4 Factors potentially impacting on the allocation of consciousness to different realities (i.e., determinants of vectorial decisions)

Determinants of vectorial decisions in the multiverse	Description	Impact on decisions	Practical considerations
Unconscious motives of the decision maker	Preferences within the person, but unknown to him, hence not explicit part of decision models (see Chaps. 9 and 11)	Potentially substantial, if the unconscious drivers are stronger than those that are consciously understood; undermining seemingly rational judgements	The more the unconscious motives are known to the decision maker, the more they can explicitly be integrated into decision models (or those models 'debiased')
Unknown motives of other decision makers (conscious and unconscious) [The question as to what extent animals are involved, here, will not be addressed in this book; my personal hypothesis, however, is that they are]	Restricting the allocation of consciousness to certain realities when operating close to the fringe of a consciousness cluster (see Chaps. 4, 7 and 10)	Most substantial when preferences are in conflict with the restrictions; undermining seemingly rational judgements	Understanding 'mainstream' and deviating opinions in minds clusters or the structure of interaction may make this an explicit part of decision models in certain cases (example of simultaneous entry games in Chap. 10)
Impact of other, including 'non-human' factors	Integration of macro-world probabilities; mostly conscious activity (see above, this chapter)	Potentially high, informing about chances and risks associated with different alternatives	Application of regular tools from operations research, forecasting etc.
Utility interdependencies between different versions of the individual	Potential variety seeking across versions of the individual, maximizing differences between experiences (versus concentrating consciousness on a few, related realities) (see Chap. 9)	Hard to judge since experiences of other versions are unknown to us	Careful deliberation about extreme versus rather moderate alternatives might have an impact on the distribution of consciousness
Decisional conflicts	Related to utility interdependencies; extreme alternatives (with large advantages in one, large disadvantages in other dimensions) might induce 'split consciousness'	Potentially substantial impact, although other versions are unknown to us	Careful deliberation about extreme versus rather moderate alternatives might have an impact on the distribution of consciousness

(continued)

Table 8.4 (continued)

Determinants of vectorial decisions in the multiverse	Description	Impact on decisions	Practical considerations
Free will	Deciding on the emphasis put on different realities within consciousness (see Chaps. 4 and 6) —final implementation of a vectorial choice, however, is not experienced, per definition, within consciousness of each version of the individual	Impact of free will higher, the larger the knowledge of the above factors (and the smaller, therefore, the impact of the unconscious and other, potentially uncontrolled factors)	Free will enhanced with a high degree of deliberation and usage of decision models as aids; enhanced also with knowledge about other 'drivers'

psychoanalysis, will be analyzed based on the impact of unconscious motives in the multiverse in Chap. 11. Unknown motives of other decision makers are most relevant when the individual is located at the edge of a minds cluster and considers decisions that bring part of the versions even closer to that edge, into situations that are characterized by very low consciousness. Since those situations are to be avoided, several alternatives might be (almost) excluded from the set of possible choices, at least center consciousness might get under some 'pressure' not to allocate primary consciousness to realities that 'move the light cone' (in its simplified, homogenous version; see Chap. 4) even closer to the edge and some (less, but still sufficiently conscious) versions right into very-low consciousness situations; I have supposed (in Chap. 4), for theoretical reasons, that a minimal amount of consciousness will still be allocated to those 'edge' or 'shadow realities,' but the amount of consciousness it too small to accommodate the version of an individual that still possesses a fair amount of consciousness. (Sure enough, some reallocation of consciousness—otherwise not explicitly dealt with here—away from such versions would be a 'temporary' fix; but eventually the individual would have to pay attention to an increasing closeness to the edge of the cluster.) Chapter 7 demonstrated that even the individual's responsibility might be influenced by this factor in certain decision situations (even though the possibility of consciousness reallocations was also mentioned there as a partial relief, in turn partially enhancing responsibility). The impact of other factors comprises normal, probabilistic considerations about future developments relevant for the decision to be made. In fact, this is the only factor of the plethora of factors listed in Table 8.4 that is looked at within standard decision models within a singular-universe perspective. Utility interdependencies and decisional conflicts both directly address potential 'relationships' between versions of the individual in different realities. They assume that the utility generated with a specific version of the individual might not completely be independent of what a different version of an individual experiences or that there is some overall utility for the individual, generated via the 'mix' of experiences that different versions of the individual make. In any case, they assume an interdependence on the level of consciousness that is not to be found within physics since versions of the individual, located within different realities, are not able to communicate or interact using any 'classical' means (for a more detailed analysis see Chap. 9). Finally, free will is as weak or powerful in a certain decision situation with specific individuals as the other factors permit.

As said, Table 8.4 also offers some practical considerations; although I have to admit that most of those are quite vague at this point. Some of those considerations will get clearer within the next section and within the remaining chapters, however.

Whereas the potential advantages of the effectuation framework within a multiverse perspective are analyzed in the next section, I should remind the reader of the fact that different concepts of utility—including further speculations on potential variety seeking between different versions of an individual as well as multidimensional utility frameworks are part of the next chapter.

Strengths of the Effectuation Framework for Usage and Further Development Within the Clustered-Minds Multiverse

Given the considerations in the last section, it is quite clear that none of the existing decision models, (subjective) expected utility theory, behavioral decision theory (such as prospect theory), quantum decision making in the form proposed by other authors (just implementing the quantum formalism), or effectuation are able to fully capture the complexity of decision making in the clustered-minds multiverse, no matter whether a normative or a descriptive perspective is taken. Vectorial choices, the core of quantum decision making according to the clustered-minds multiverse proposed in this book, have just not been taken care of in the development of any of the approaches—although the formal framework shares some similarities with certain multidimensional utility models (see the next chapter). It is clear that only a long-term research endeavor towards the development of a novel, quantum decision making concept might suffice (see also the final discussion and conclusions section). But is there any decision model whose adaptation might be quicker, that is 'better equipped' as a starting point for such an endeavor than others? (Clearly, that is not to say that this will *eventually* be the best approach.) Also, it might be asking for too much generating 'clean' normative and descriptive approaches at this point—pragmatic decision aids might suffice.

The most flexible approach, well-suited for high-complexity situations, is certainly effectuation. And I have mentioned, within the list of premises, that utility considerations per se might be important but *maximization* with the goal of finding 'the' optimal alternative is not an easy match with the vectorial decisions to be made in the clustered-minds multiverse, unless the exact utility interdependencies between different realities as well as restrictions for allocations (e.g., the requirement to choose moderate allocations) are better understood. Therefore, I am going to look at the 'development potential' of the effectuation approach as a decision aid, comprising descriptive as well as normative aspects, more closely and walk through all the criteria, listed in Table 8.4, applying an effectuation perspective. The result is to be found in Table 8.5, having the 'determinants of vectorial decisions' in the first column, 'effectuation as a decision aid' in the second column and 'theoretical perspectives,' possibilities to further develop or support the effectuation approach, in the third column. I am not claiming that Table 8.5 is using effectuation directly; the idea is to look at perspectives in the 'spirit' of the effectuation approach. As was the case with Table 8.4, the moderate allocation of consciousness requirement as well as the specific situation for reallocations of consciousness, are left out of the explicit analysis. However, it is anyway unclear whether effectuation is as much of an appropriate model for consciousness reallocations as it is for 'primary' vectorial choices.[16]

Looking at Table 8.5, three aspects become crystal clear. First, looking at the determinants of vectorial choice in the spirit of the effectuation approach leads to a rich paradigm, it generates a picture of real decision making that is consistent with the clustered-minds multiverse and that is of some practical value, already. Second, the table reveals how complex real-life decision situations typically are, even though we are not normally aware of it. On the one hand, somebody trying to come up with a new *formal structure* for the decision sciences based on the clustered-minds multiverse might easily reach a state of even deeper frustration than that already reached, perhaps, much earlier in this chapter when looking at the plethora of factors involved. On the other hand, that might just imply that a lot of effort needs to be invested into simplifications or abstractions—somewhat difficult at this stage of development, however. Third, research opportunities are overwhelming. Regarding partially known unconscious motives, the development of a 'tool box' of self-binding commitments is a potentially exciting research field. With self-binding commitments, according to Schelling (1981 [1960], 1978), a decision maker purposely limits the scope of decisions as well as their potential outcomes. An example is an individual that wants to stop smoking. The most important self-binding commitment is to not have any cigarettes available in the house.

Regarding the unknown motives of others, relevant when operating at the edge of consciousness clusters, certain communication strategies have to be looked at and applied in a fairly different way than usual (they have to be applied in an ethical way, too, since the mere fact that *we* would like the minds cluster to move in a certain direction is not a sufficient justification for the respective activities). Regarding the impact of 'other' factors, it is quite clear to me that we possess more scientific knowledge about this determinant of decisions than about the others, but it is hard for me to see what kinds of challenges might arise here as well when

[16]Answering this fairly complex question is beyond the scope of this book.

Table 8.5 Applying the 'spirit' of effectuation to vectorial decisions

Determinants of vectorial decisions in the multiverse	'Spirit of effectuation' as a basis of decision aids	Theoretical development perspectives
Unconscious motives of the decision maker	• Safeguarding against the allocation of consciousness to potentially harmful realities, driven by unconscious motives; in case such motives are partially known, *implementing self-binding commitments* (Schelling 1981 [1960], 1978)	• Better understanding changes in allocation of consciousness when unconscious motives become conscious • Better understanding and developing self-binding-commitment strategies for different scenarios
Unknown motives of other decision makers, conscious and unconscious	• Impossible to know all those motives, but partially possible to influence them • Personal communication with 'closer' individuals • 'Anonymous' communication, e.g., via social media, with more 'distant' people	• Better understanding the impact of different types of communication on the alignment of different individuals' realities • Developing prediction techniques for 'movements' of minds clusters
Impact of other, including 'non-human' factors	• Making small steps; staying flexible to see how things develop and to react later • Trial-and-error approaches rather than 'big jumps' • Affordable-loss principle and satisficing rather than optimization etc.	• Application and further development of the 'tool box' already proposed within the effectuation approach
Utility interdependencies between different versions of the individual	–	–
Decisional conflicts	• Stepwise approach avoiding extreme tradeoffs and allocation of consciousness to extreme alternatives such as high-risk, high-return profiles	• Simulations of different outcomes with different weights on attributes of alternatives (see also Chap. 9 on multi-criteria models) • Graphical visualization

(continued)

Table 8.5 (continued)

Determinants of vectorial decisions in the multiverse	'Spirit of effectuation' as a basis of decision aids	Theoretical development perspectives
Free will	• Stepwise approach leads to many small decisions where in each of them, free will may play out • Effectuation generates more feedback on previous choices so that learning is enhanced, increasing the chance of learning about the other determinants	• Understanding free-will enhancing activities requires further developing a science of consciousness [It is clear that I am not the first one to propose the need for a science of consciousness; indeed, the Toward-a-Science-of-Consciousness (TSC) conference is a clear sign that the need is seen by many; however, I would like to argue that the clustered-minds multiverse injects a larger need for quick progress into this endeavor, it makes it *inevitable* to proceed this way—analyzing the impact of consciousness on vectorial choices might constitute an important subfield in this area] • Is, e.g., free will enhanced with many small, iterative choices in the spirit of effectuation (as supposed, here) or rather a few big ones, more consistent with causation?

research within the framework of the clustered-minds multiverse expands. The trickiest issue are the utility interdependencies between different versions of the individual. As easy to see from Table 8.5, I had nothing to say at this point, from the perspective of effectuation, but I have to admit that I see this as a generally difficult, highly speculative domain (at least I was unable to come up with more than already stated in Table 8.4, where my 'practical considerations' were the same as with the 'decisional conflicts'). A feel as to what might be learned and understood in future research, however, could perhaps be gained based on a more detailed discussion of potential utility independencies in the next chapter. Most directly connected, I suppose, to the effectuation approach, are the analyses and recommendations for decisional conflicts and free will. Both domains appear to benefit from the stepwise, the iterative, the trial-and-error approach suggested within effectuation.

Summary, Final Discussion, and Conclusions

This chapter started with a section on classical decision theory, the so-called normative or standard expected utility theory, including its subjective expected utility variant. It went over to behavioral decision theory in the next section, followed by a section on the novel perspectives offered by the effectuation framework and then, in the subsequent core section of this chapter, analyzed the problem of vectorial decision making, a radically different setup than in a singular universe, required within the clustered-minds multiverse. It is evident from that section that the development of a new *normative* and *formal* framework for the decision sciences requires a large amount of further research as a prerequisite, because of, first of all, the large difference between the selection of a singular alternative in standard decision theory and vectorial choice in quantum decision making as well as, secondly, the fairly complex set of premises formulated there for quantum decision sciences. (Note that a clear-cut descriptive account is, anyway, difficult if a normative benchmark is missing.) I would like to argue, however, that the modelling of the dependent variable: a vector of degrees of consciousness allocated to an individual's versions (most probably an appropriate definition of the objective function, if utility interdependencies between versions as well as restrictions for allocations could be specified in a satisfactory way) is slightly closer to 'completion' than the modelling of the independent variables (restrictions and drivers of vectorial choice). What do I want to exactly express with this statement?

The formal definition of vectorial choice (with example) in Formulas (8.1) and (8.2) is an important step, already, to the dependent variable of a model of quantum decision making, although the specification of the overall utility for an individual from different vectors of degrees of consciousness allocated to different versions awaits intensive future research. Part of what was listed in Tables 8.4 and 8.5 potentially contributes to the understanding of that utility, albeit in a fairly imprecise way: utility interdependencies between different versions of the individual as well as decisional conflicts could be used, if better understood, to derive the *overall utility* of a vectorial decision over all versions of an individual. Another restriction, left out from Tables 8.4 and 8.5 for the sake of simplicity, the requirement of implementing moderate allocations of consciousness (see the list of premises of the clustered-minds multiverse in Chap. 4), would also have to be taken into account. Moreover, triggers and restrictions etc. of reallocations of consciousness (see Box 6.3 in Chap. 6) would have to be better understood. Most of this might, for a longer period, be theoretical and philosophical research until, probably seen as difficult to impossible by many at this point in time, certain empirical research on distributed consciousness might be possible. E.g., an indirect test of utility interdependencies is conceivable, based on the happiness of the version residing in our joint reality, after different (experimental) interventions that would, according to a then hopefully better developed theory, have certain predicted effects, via utility interdependence with versions in other realities, on the version in this reality. A major challenge would then be to experimentally control for alternative explanations of changes in happiness due to, say, *cognitive dissonance* (Festinger 1957) with one version of the individual. In addition to this (and very helpful for this type of experiment) one might try to directly measure the degree/strength of consciousness of a person (in this context: the version of an individual residing in our joint reality), perhaps via known techniques such as an *electroencephalogram* (EEG),[17] via BOLD (Blood Oxygen Level Dependent) contrasts achieved using MRI or perhaps via newly developed measurements.

Some of the restrictions and drivers of such a vectorial decision, i.e., part of the other determinants listed in Tables 8.4 and 8.5, are, however, even harder to model. How would, e.g., unconscious motives (potentially influencing all seemingly rational considerations of a decision maker) or unknown motives of other decision makers, be appropriately modelled? Free will per se does not have to be modelled, I suppose, since it is the impact an individual's consciousness has on vectorial decisions and thus the *'entity' to be informed* by decision models. What might be interesting here, however, would be to shed more light on the blackbox of the transition from 'free-will deliberations,' that are conscious, to 'free-will implementation' in the form of a vectorial choice that is not taking place under conscious control (i.e., there are no qualia for the event of splitting minds). Finally, the standard decision-theoretic toolbox is quite appropriate for grasping the lack of knowledge of future developments (i.e., the macro world probabilities as expressions of that lack of knowledge) a decision maker is typically faced with (within a singular world or within multiple realities, for that matter).

The previous section then dealt with the potential advantages of the effectuation framework—and further developments based on this approach—for usage as a decision aid within the clustered-minds multiverse. In a way, further developments based on the effectuation approach turned out to be, altogether, the simpler route to be taken, at least short-term, compared with the development of formal normative and descriptive frameworks of decision making. Following this premise, the previous section explored the potential of formulating aids for dealing with most factors listed in Table 8.4, using effectuation within the situation posed by the clustered-minds multiverse. Based on this analysis, effectuation and its possible extensions/modifications appear as a practical approach towards decision making in the multiverse that smartly deals with complexity (revisit Table 8.5 and the discussion based on it). A major driver of that complexity (on top of the complexity already indicated by the numerous factors listed in Tables 8.4 and 8.5 and factors even left out of consideration within those tables for the sake of slightly simplifying

---

[17] I have no sound reason to suppose that EEG measurements might be the right way to go in the context analyzed here, but intuitively I felt that this is the closest to an appropriate measurement to try out that we currently possess. Although brain waves have never been quite understood, close to a study of consciousness, early results achieved with the EEG already indicated the following: "(…) [The] state of wakefulness and sleep of a normal individual (…) has been related successfully to changes in the EEG" (Simon and Emmons 1956, 1066). Moreover, depth of sleep has been related to certain patterns in EEG, and alpha waves are seen as an index of consciousness (Simon and Emmons 1956, 1066).

the picture, e.g., being restricted to moderate allocations of consciousness) has already been mentioned, above, and will appear more clearly in the next chapter: A vectorial choice between *realities* typically resembles a real-options structure (even though partially countered by reallocations of consciousness; see Box 6.3). And effectuation, in combination with the appropriate utility concepts (see the next chapter), might also offer a possibility to smartly reduce the complexity of that part of the problem.

Many non-behavioral economists, cognitive psychologists and other researchers interested in normative decision benchmarks, however, will find the (perhaps temporary) move towards the effectuation approach and the (at least temporary) wait for the completion of a formal framework of decision making that accommodates for the new interpretation of quantum mechanics proposed in this book unappealing. They might argue that unless they are presented with a complete decision *benchmark*, they are unable to understand rational decision making implied by the clustered-minds multiverse. In fact, there is nothing I can answer beyond what I have already said: Since effectuation is not only a framework that is somehow able, with some adaptations, to deal with the multiverse, but also a totally different perspective on decision making in general (also in a singular universe), the argument that effectuation might perhaps 'work better' than any novel, formal (and normative) framework of decision making in the clustered-minds multiverse is speculative. It is also speculative (albeit plausible) to assume that modifications of the effectuation framework are indeed 'quicker' than the development of a new normative benchmark, suited for the conditions in the clustered-minds multiverse.

Moreover, the chapter turned out to be almost blank on the relationship between existing approaches of behavioral decision making and the clustered-minds multiverse so far. Could, on the one hand, anything be learned from the conditions relevant in the clustered-minds multiverse for the understanding of some of the behavioral phenomena? Or would, on the other hand, the understanding of such behavioral phenomena better inform the decision maker on the determinants of vectorial choices, hence adding more aspects to Tables 8.4 and 8.5 or understanding some of the aspects already listed in those tables (or their practical handling) somewhat better? How would, e.g., the persistent finding, modelled within prospect theory (e.g., Kahneman and Tversky 1979), of the reference dependence of choices be located within a multiverse framework? Is reference dependence, together with the principle of *loss aversion*, resulting from a comparison of positive and negative changes relative to some reference point (e.g., Tversky and Kahneman 1991), perhaps weakly related to self-binding commitments (Schelling 1981 [1960], 1978) that have been proposed to deal with, e.g., the influence of the unconscious on vectorial choices, in the sense that *precautions* should be implemented excluding, as much as possible, realities from the spectrum of developments where heavy losses could be encountered? It is quite clear that these considerations are also related to the affordable loss principle from effectuation, that has been listed, already, in connection with the impact of 'other factors' in Table 8.5, so that indirectly, aspects of behavioral decision making have been linked to the discussion, already (albeit rather by Sarasvathy 2001, linking behavioral decision making and effectuation in this part, than by myself); it is also quite clear that the brief discussion in this paragraph is not more than just a start and that the further exploration of the relationship between behavioral decision making and the clustered-minds multiverse will turn out to be fruitful.

Finally, there is the 'smart-and-frugal heuristics' program by Gigerenzer and coauthors. This approach should be considered an important part of the decision sciences, a central part of behavioral decision making. Proposing the smart use of certain heuristics to be superior to optimization in contexts where individuals simply cannot do a good job in this regard (see, e.g., Gigerenzer et al. 1999) exhibits some proximity to the effectuation approach. Future research might want to explore closer connections between the clustered-minds multiverse and Gigerenzer et al.'s approach to decision making: Are there any heuristics that work so well because they are smart, given the situation of a clustered-minds multiverse? Does the clustered-minds multiverse perhaps even allow for a better theoretical justification of those heuristics than a singular universe?

As already stated at the beginning of this chapter, two more chapters on decision making follow. They will offer additional insights on decision making in the multiverse on the one hand, but they will partially add to the complexity arising in this chapter rather than resolving it on the other hand. So, the spectrum of approaches discussed in this chapter as well as in Chaps. 9 and 10 might inspire thoughts as well as future research; they might demonstrate how exciting the multiverse analysis turns out to be for the decision sciences. The reader will often get a clear idea as to what will have to be revised or even abandoned with respect to the decision sciences, both from a normative as well as from a behavioral standpoint, but he will sometimes get no clear idea as to how exactly those changes will have to look like or by what exactly those abandoned parts will have to be replaced. It is in the nature of things that an approach as radical as the clustered-minds multiverse suggests many changes, and that those changes might require a lot of time and research efforts to be made.

Chapter 9
Different Concepts of Utility

Existing Utility Concepts Are (Very) Different: Which One (If Any) Works in the Clustered-Minds Multiverse?

Utility is the major ingredient of individual decision models, and this completely holds in vectorial choice; it is also one of the main drivers of reallocations of consciousness. The concept of utility has already been touched a few times in the last chapter, and clearly at some important points (see, e.g., Tables 8.4 and 8.5). However, a detailed analysis was missing in that chapter and potential problems with different utility concepts were not expounded. The current chapter now concentrates on the notion of utility and discusses, which (if any) of the existing utility concepts might make sense in the clustered-minds multiverse. Specifically, is there a concept of utility that helps understanding how much weight some individual puts on different realities in vectorial choice?

In the last chapter, I have traced back the idea of utility to Bernoulli (1954 [1738]), but others might rather associate the idea of utility with Bentham (1996 [1789]) or see it as even older than the concepts of both Bernoulli and Bentham (for excellent analyses of the development of utilitarianism see, e.g., Kauder 1953; van Daal 1996). However, Bernoulli's (1954 [1738]) concept clearly foreshadowed the concept of *decision utility* that is the core concept in normative decision theory and that has been the dominant concept in *mainstream economics* and many other applications in the 20th century.[1] Whereas Jeremy Bentham (1996 [1789]) developed a concept of *experienced utility*. The importance of the distinction of decision utility and experienced utility has been emphasized by Kahneman et al. (1997), who criticize the dominant usage of the concept of decision utility and propose to *replace* it with the concept of *experienced utility* in the decision sciences as well as

[1]Many economists would argue that this still is the central paradigm in economics. Whereas a growing group of behavioral economists would disagree.

© Springer Nature Switzerland AG 2018

C. D. Schade, *Free Will and Consciousness in the Multiverse*,
https://doi.org/10.1007/978-3-030-03583-9_9

in economics. What are the differences between those two concepts? Is one of the two preferable, from the perspective of the clustered-minds multiverse? And, given the—perhaps temporary—importance of the effectuation framework for decision making in the clustered-minds multiverse according to the analysis in the last chapter, is there a possibility to integrate any of them into the concept of effectuation?

In the next section, I will discuss whether Kahneman et al.'s proposal makes sense or whether experienced utility rather solves some problems whilst creating others. What are the advantages and disadvantages of decision utility and experienced utility? I am going to look at this quite generally as well as from the perspective of vectorial choice. As it will turn out, decision utility in its existing form, despite important merits, is insufficient for usage within the novel framework proposed in this book. But it will turn out that—among other problems—the fact that individuals might sometimes have *unconscious dispositions* for making odd and painful experiences in their lives makes it also difficult to use experienced utility theory in its current form. The subsequent section will intensively deal with that problem and tries to better understand as to *why* people might at all look for odd and painful experiences—an aspect that also plays a role in Chap. 11. To do so, I will evaluate the psychological literature on violent entertainment, e.g., literature on why people are watching violent movies, hence the 'metaphor' of consciousness putting a certain emphasis on different 'movies' (i.e., certain versions of reality) will be taken literally, here (keeping in mind the limitations this has for understanding 'real'/outside cinema behavior). I will also look at results from consumer behavior relevant for this discussion. Another section then addresses again, and in more detail, the question whether there could indeed be *interdependencies* between the experiences and the utilities of different *versions* of one individual, as has already been hypothesized and discussed in the last chapter (see, e.g., Tables 8.4 and 8.5). Do I, e.g., suffer more when other versions of mine suffer; and am I happier if other versions are happier? Is there an unconscious 'bridge' between one version of an individual and another? Or is there even an *overall concept* of making experiences that implicitly takes into account the states of all versions of one individual, distributing and optimizing experiences for the entire 'group' of 'versions?' Coming back to the odd and painful experiences, does the explanation for this phenomenon perhaps partially lie here? Another section deals with the question as to what concept of utility might be used within the framework of effectuation. Here, the fact that the utilities of different selected realities are playing out long-term and determining them within a normative framework would often lead into a real-options structure that would—in narrow terms—even require the usage of a different decision model (a modification of the normative approach introduced in Chap. 8; the theory has been introduced by Dixit 1992; Dixit and Pyndick 1994), leading—in combination with the multiverse—to an extremely complex setup, has to be taken into account as another advantage of the effectuation approach, if combined with an appropriate utility concept. The chapter closes with a thorough discussion of the problems in current utility analysis and what might be the next steps in developing utility analysis consistent with the clustered-minds multiverse, i.e., with vectorial choice. I will thereby consider what might be learned in future research from integrating approaches that are looking at multidimensional utility, partially informing the issues that have been raised around decisional conflicts in Tables 8.4 and 8.5 and following those tables in the last chapter.

"Back to Bentham?" Are There Advantages of Experienced Utility over Decision Utility from the Perspective of the Multiverse?

What Is Different with Experienced Utility?

As already mentioned, in their seminal paper with the title "Back to Bentham: Explorations of Experienced Utility," Kahneman et al. (1997) propose to move away from *decision utility* (see also Read 2007; Kahneman 2000; Kahneman and Sugden 2005). But what exactly is decision utility, and what are the pros and cons of it within the perspective of this book? Decision utility reflects the standard decision scientist' and economist' idea that *preferences* can be inferred from *choices* (rational choice theory): "In modern decision research (…) the utility of outcomes refers to their weight in decisions: utility is inferred from observed choices and is in turn used to explain choices" (Kahneman 2000, 673). In other words, "choices provide all necessary information about the utility of outcomes because rational agents who wish to do so will optimize their hedonic experience" (Kahneman et al. 1997, 375).

From the perspective of a *rational choice* theorist, this position has huge advantages. Here, I shall mention only two: One is that '*speculation*' about inner states of a person is seen as unnecessary; the other is that a discussion as to *why* a person chooses something, whether this *appears* reasonable or not, is viewed as unnecessary. If somebody smokes, e.g., this maximizes his utility. If somebody engages in risky sports, fine. It reflects his preference. More generally, if someone opts in favor of odd, even painful experiences, this is nothing the researcher has to worry about. In the first section of the last chapter, another advantage of the most influential rational choice model, expected utility theory, was explained: the possibility of integrating *risk preferences* of the decision makers.

The big disadvantage of this concept (disregarding additional problems that arise within the multiverse for a moment; but see the discussion in the next section) is the mere fact that *rationality* has to be *assumed*, and rationality has been shown not to hold with real decision makers (see e.g., Kahneman and Tversky 1979; Tversky and Kahneman 1974).[2] What we observe as the choice of an individual might not at all reflect the preferences of this individual but rather his errors, his biases, his unconscious desires etc. Perhaps, the smoker only smokes because of a strong

---

[2]Note that this objection cannot be countered by the argument that a normative analysis was claimed to be impossible at this point, anyway, in the last chapter. Rationality here is part of the *definition* of this utility concept.

addiction? And the risky sports activities try to compensate an inferiority complex that should better be healed otherwise? Or, choosing a risky option in roulette may not result from a risk-loving preference but rather from the fact that the player has just lost money and tries to break even (Thaler and Johnson 1990). If situations like this are prevalent (and they are!), then the idea that decision makers maximize utility with all their chosen activities is impossible to hold. Consequently, the assumption of preferences being revealed in choices is *incorrect*.

This lead Kahneman et al. (1997) to postulate a new perspective on utility supposed to replace rational choice theory: *experienced utility* theory. Kahneman et al. (1997) state that experienced utility can directly be *measured* and does not have to be inferred from choices. Their theory is based on earlier thoughts by Bentham 1996 [1789] and Edgeworth 1967 [1881], and it looks at utility explicitly as something that maps the pleasure and pain arising from the experience and that accrues to an individual *over time*.[3] According to Read, "Jeremy Bentham's moral philosophy centered on three claims: the goodness or badness of experience is the pleasure or pain arising from the experience, this pleasure or pain is (in principle) quantifiable, and the quantities so obtained can be added across people" (Read 2007, 46). The latter implies the ideal case of a *cardinal utility*. And from Edgeworth (1967 [1881]) one can derive the basic idea of a *moment-utility function* that can be used as an appropriate formal framework for Bentham's utility concept for this ideal case. Note that a cardinal utility implies that there is a *natural zero point*, and that the measurements are *objective*, such as with the measurement of *length*. Cardinal utility is often seen as a *purely hypothetical* case, e.g., because experiences are *reference-dependent* (e.g., Kahneman and Tversky 1979). The claim of a cardinal utility is also hard to hold empirically, given the limitations in measuring experienced utility.[4] However, applying this ideal theory, the total utility of watching the movie would equal the sum of utilities the individual experiences, say, each minute; mathematically more elegant would be to assume that the cardinal utility accrues *continuously*.[5] According to Kahneman (2000), indeed building up directly on Edgeworth 1967 [1881]: "With cardinal measurement, the most natural index of total utility could be calculated: the temporal integral of moment-utility" (Kahneman 2000, 680). The idea is simple and depicted in Fig. 9.1. Mathematically, the integral of the moment-utility function at the top of the figure would have to be calculated between boundaries t_1 and t_2.

---

[3]It is clear that time is used by the authors in psychology and economics in its traditional meaning —an objective flow of events in one direction. Whereas when I am using and analyzing the notion of time, I mean to interpret it in the subjective sense as an experienced flow of time. Fortunately, this 'mix' of interpretations does not lead to any problems throughout the chapter. Indeed, one can read the psychologists' and economists' reasoning as if they were talking about subjective time. Remember that objectively, as pointed out in Chap. 3, there is no flow of time but only different parallel realities.

[4]As shown below, Kahneman et al. (1997) concentrate on the ordinal interpretation of utility (where only rankings of different experiences are required from a decision maker).

[5]Here, the question is how frequent an individual updates his state of utility. A continuous measurement is probably as hypothetical as cardinal utility is.

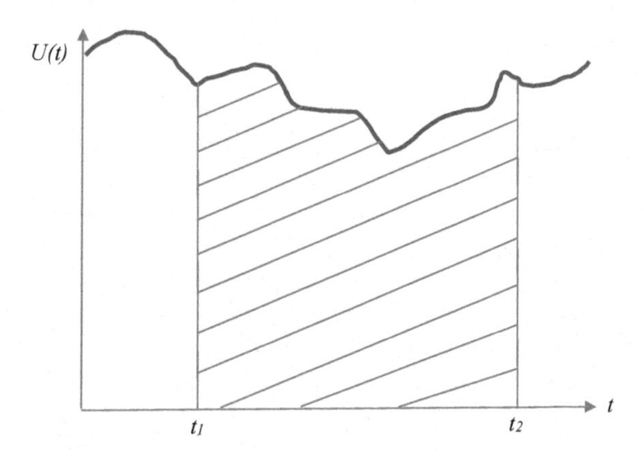

Fig. 9.1 Total utility of watching a movie in a cinema with t_1 = start of the movie and t_2 = end of the movie; the shaded area indicates the total utility of the visit (based on Kahneman et al. 1997; Kahneman 2000; in turn based on Bentham 1996 [1789]; Edgeworth 1967 [1881])

Experienced Utility or Decision Utility in the Clustered-Minds Multiverse—What Is the Appropriate Concept to Be Used in Vectorial Choice?

Up to here, I have been quiet about the requirements of the clustered-minds multiverse with respect to the appropriate utility concept. But it seems evident that the example depicted in Fig. 9.1 nicely matches with the experience of an ongoing sequence of events in one reality or with one version of the individual, i.e., with consciousness navigating the clustered-minds multiverse and making snapshots of different realities (including those that represent different points in time) where each of them generates a different utility. Theoretically, there is no difference between the *measurement* of a total utility of a movie watched in a cinema and the total utility of some other, specific reality experienced during some activities at home or during a subway ride downtown; the difference—and that makes matters a bit more complex outside the cinema—is that outside the cinema, individuals can change the movie within parts of a second. Experienced utility appropriately grasps the effects of the 'split-second-type' flow of experiences of an individual navigating the multiverse. Whereas the rational-choice-type idea of decision utility appears somewhat anemic in this context. This is not to say that rational choice theory could not in principle be turned into some moment-utility form, too. One of the remaining problems here would be the exact measurement of decision utility at different points in time since *defining and measuring choices quasi-continuously* is even harder to imagine than a quasi-continuous measurement of experienced utility.[6] The other is

[6]However, choices made in split seconds are not only *a realistic case* but there is simply no other way as making those choices all the time (I am grateful to a discussion with Tanja Schade-Strohm where this became clear to me). So, in terms of pure theory, decision utility, assumed to be relevant for each split-second decision, would offer a good framework if people could be assumed to make rational choices. However, this assumption is incorrect and the practical definition of a novel decision situation at each split of a second is hard to imagine.

the real-options structure of subsequent decisions. The decision to put a lot of emphasis on one rather than another reality might partially determine the next vectorial choice, e.g., when already operating close to the fringe of a minds cluster, or when the decision *removes* an option (say, sell an antique item for a middle-level price) that might, after certain developments, have a much higher market evaluation and cannot be sold for that price, anymore, because it is not still in the possession of the individual.

So, from the perspective of usage within the framework of the clustered-minds multiverse, it seems as if experienced utility were 'ahead' of decision utility at this point. In its current version, it seems more appropriate for usage within this novel framework, and decision utility additionally suffers from the fact that people are not rational. Partially, however, this relatively positive evaluation of experienced utility might be, however, a result of an implicit simplification. Removing the postulate of rational choice seems to remove the necessity to take into account the real-options structure of many decisions, now described by experienced utility. It shifts this part into the consciousness of the decision-making individual that now must find a—perhaps simplifying—way to deal with that feature of subsequent decisions. Thus, there are reasons to believe that none of those two concepts suffices in its current form unless an appropriate framework for decision making in the multiverse has been crafted; this thought will be extended in the last section of this chapter.

But let me now turn to some technical 'details,' that are important for dealing with experienced utility. Whereas *cardinal* measurement of experienced utility might be a substantial challenge, continuous measurement can at least be seen as a *mathematical idealization* of subjective utility updated, say, every minute (a measurement requiring reports by the individual is hard to imaging within intervals of below ten or twenty seconds); talking about one minute should be considered 'safe.' And Kahneman et al. (1997) built up an *axiom system* for this new theory of decision making that closely resembles the above thoughts on moment utility but only requires *ordinal measurement*, i.e., individuals only have to be able to *rank-order* experiences in terms of pleasantness (Kahneman et al. 1997, 389); furthermore, that axiom system catches Edgeworth's intuition depicted in Fig. 9.1 as well as the original idea by Bentham 1996 [1789].

There are nevertheless three potential problems with experienced utility theory that might make it difficult to use it for the analysis of vectorial choice. The first is the fact that, as is the case with multidimensional utility approaches (see the last section of the current chapter), integrating risk preferences (e.g., risk aversion) of a specific individual into his experienced utility might be a challenge. A possible solution might be to integrate risk preferences into a risk-weighting function, however, in the spirit of cumulative prospect theory (Tversky and Kahneman 1992) and to leave the utility component untouched. But this leads to the question whether quantum decision making will *ever* be more than a *decision aid* with some plausible, formal background—whether or not a normative/prescriptive theory is even *conceivable.*[7] The second, to be analyzed in the following, is the fact that so-called *remembered utility*, measured after the respective event (and important for subsequent choices) strongly deviates from experienced utility measured 'real-time.' The third, to be analyzed in more detail afterwards, is the fact that many people clearly *choose* in favor of *odd and painful experiences* at least sometimes and that it is hard to make sense of this behavior in terms of experienced utility.

Let me now carefully look at the issue of remembered utility. In a remarkable paper by Redelmeier et al. (2003), 682 patients undergoing *colonoscopy* participated in a randomized experiment where their experienced and remembered pain was measured. Reconsidering the logic explained in Fig. 9.1 but now considering the utility in this graph to be a negative one (such as the pain from colonoscopy), *extending* a *negative experience* by a certain *amount of time* should *increase* the integral of the negative experience (i.e., decrease the utility). So, individuals should *dislike* a negative experience that is extended by a certain amount of time *more* than a negative experience that is shorter. This prediction turns out to be correct in terms of utility measured over time (i.e., every minute). But it turns out to be incorrect in terms of *remembered utility*. Specifically, with one group (i.e., the 'modified care' group) in the experiment by Redelmeier et al. (2003) "the tip of the colonoscope was allowed to rest in the rectum for up to 3 min prior to removal (…). [M]odified care lengthened the duration of the procedure but resulted in final moments that were less painful" (188). According to Kahneman et al. (1997), there is empirical evidence from various contexts that individuals' remembered utility is generated the following way: "(…) the remembered utility of pleasant or unpleasant episodes is accurately predicted by averaging the peak (most intense value) of instant utility (or disutility) recorded during an episode and the instant utility recorded near the end of the experience"[8] (381). Thus, it was expected that in the colonoscopy experiment the patients with the longer negative experience would have a more favorable memory because the negative experience at the end of the procedure was less negative than the utility at the end of the experience in the other experimental group that did not experience the prolonged procedure. The predictions of this so-called *peak-end rule* were fully confirmed. Thus, respondents that underwent the *longer* procedure, leading to a *higher* sum of negative instant utility, were *more* likely to choose in favor of another colonoscopy in the future (Kahneman et al. 1997; Redelmeier et al. 2003).

What does this imply for a theory of utility that works within the framework of the clustered-minds multiverse? The consequence is hard to determine, indeed.

---

[7]Using a risk-weighting function to capture risk preferences of the individual leaves the formal framework of rational choice, because it cannot be derived from the rationality axioms of EUT. Note, however, that certain modifications/generalizations of the axiom system do allow for risk weighting. A detailed discussion is beyond the scope of this book.

[8]Note that this is one of the occasions where something similar to *reference dependence* is integrated into the analysis of experienced utility.

The task is to finally understand *how exactly* individuals experience (and thus choose in favor of) certain weights within a set of different realities, i.e., how they experience the consequence of a vectorial choice. It is unclear what the more appropriate definition of experienced utility is in this context: utility that is measured real-time, i.e., from moment to moment, or remembered utility? Perhaps I should remind the reader of the fact that the explanation of free will offered in this book partially rests upon the notion of consciousness influencing our brain reactions backwards (see Chap. 3), or more precisely, using the framework of vectorial choice, what type of reality, in terms of a connection between brain reactions and 'later' observed choices, it allocates more weight to? Let me suppose that utility eventually pertains to the domain of consciousness or qualia (where else?) and that future and past, memory and foresight turn out to be very similar in this domain just because consciousness navigates a multiverse with parallel times (derived from Chap. 3); this applies even though perceptual conventions might make it much easier to look backwards than forwards. If individuals where then to (perhaps only partially) '*anticipate' remembered utility* and would try to maximize this, one would have to put the peak-end rule in the middle of the analysis. If one, however, assumes that the flow of experiences over time, the 'snapshots of reality' taken (see Chap. 3) and the respective utilities experienced, are essential,—in this case kind of a 'myopic' behavior—and then also disregarding the potential real-options structure of the decisions and solely relying, as the then only remaining possibility of corrections, on reallocations of consciousness, the analysis would have to look differently. I will come back to this problem in the final discussion section of this chapter.

I now turn to the issue of individuals *choosing* odd and painful experiences, i.e., allocating a major amount of consciousness to painful realities. This ranges from the special case of watching horror films (here the individual might have at least known the trailer) and repeatedly playing computer war games (where the content is known to the individual after the first time he played it or even before that) to living in a city that makes the individual unhappy. Note that there will be unconscious influences on what reality will be experienced to what extent. Thus, chances are that if an individual turns out to be unhappy, within realities with a considerable amount of consciousness allocated to them, that he might have co-created this reality somehow via *unconscious motives*. But let me assume that a person has really just been unlucky with his move to a certain city as well as with the company he is working for, one would still have to explain why he *stays* in all those unhappy situations.[9] So we do have to take seriously the fact that individuals partially *choose* to make negative experiences. The entire issue is actually more complex since on the one hand, the requirement to implement moderate allocations of consciousness

[9]Some explanations from behavioral decision theory would be escalation of commitment, psychological sunk cost, status quo bias (e.g., Kahneman et al. 1991; Samuelson and Zeckhauser 1988; Burmeister and Schade 2007; Sandri et al. 2010) etc., but those types of reference-dependent behaviors of inertia are most certainly not sufficient to fully capture the phenomenon.

and other restrictions[10] (see, e.g., Chap. 7) might make it difficult to totally (or better: almost totally) avoid certain realities altogether; on the other hand, taking into account the possibility of reallocations of consciousness, it is indeed surprising as to why people 'stay,' say, with unhappy versions of themselves, equipped with a considerable amount of consciousness.

The issue of individuals partially and perhaps unconsciously 'volunteering' for bad or unpleasant experiences in turn has two aspects. One is the question as to what could be the *reasons* of such behavior, to be discussed in the next section of this chapter. The other aspect is more of *theoretical nature*. Choosing in favor of (and continuing for decades, say) an unhappy relationship would be nothing to 'worry about' (this is not at all meant as a low empathy statement but only as a statement about theory!) within the framework of decision utility (and assuming a singular reality) since rational behavior mirroring the individual's preferences would simply be assumed—e.g., a masochistic personality where the distant and aggressive partner is *finally* preferred over a peaceful and loving relationship.[11] But how could one possibly make sense out of this in terms of experienced utility? Most certainly, when the individual is asked to report all his moment utilities from 6 a.m. to 10 p.m., say, perhaps in minute intervals, he will report a very negative total utility if he spent the day with his significant other. Over the years, the integral of negative experiences becomes larger and larger then; and shouldn't everything tell him to stop that activity, to see a counselor, to separate? Or is the idea in experienced utility that 'negative events' generate 'negative utility' even fundamentally flawed?

People Choosing Bad or 'Mixed' Experiences: What Can Be Learned from Cinema Films and Consumer Behavior?

It is obvious that there are differences between watching a cinema or TV film, engaging in video games and experiencing our lives outside those activities. Goldstein, e.g., states that the "potential of a book, film, or video game to engross one in an *imaginary* world is one of the most attractive features of entertainment

[10]One could certainly make the point that, in close analogy to the considerations in Chap. 7, the people that one meets making odd and painful experiences are always very-low-consciousness fellows, kind of 'forced' to also experience this reality with a minimal amount of consciousnes. But this argument is not plausible. First of all, why should we share realities with very-low consciousness versions of individuals when in fact we should avoid such situations? Secondly, *we* are also most probably sometimes opting for such experiences, and are *we* actually very-low-consciousness versions of ours? (Even though this can never be answered with final certainty, I simply do not hope so.)

[11]Eventually, within the framework of decision utility, one would have to assume a positive utility of staying in a bad relationship for this individual. But most certainly, that is not what the person experiencing this situation will report in terms of experienced utility.

media (…). This potential (…) helps explain the tolerance for, if not the attraction of, violent imagery" (Goldstein 1999, 275; italics mine). However, there might also be similarities with 'reality,' with the 'real-life movie' we select. In fact, both spheres share one puzzle, already in a singular world: Why are people selecting unpleasant or even violent 'experiences' at all? The plethora of potential motives that have been suggested as well as the explanations that have been put forward with respect to violent entertainment are overwhelming. A good insight can be achieved with the several contributions to the book *Why We Watch*, edited by Goldstein (1998). Sure enough, some of those motives are *specific* to films, video games and certain sports such as boxing or wrestling, but some motives might generalize to the sphere outside entertainment. The reason why I feel that a closer look into this might be helpful with addressing the question analyzed in this chapter is straightforward. People *voluntarily*[12] go to the movies or play video games. In fact, they view this as *entertainment*. So, if one wants to understand why seemingly 'negative' feelings such as fear are *experienced* in 'real-life,' without people being under outer force, results on violent or disturbing entertainment might turn out to be informative.

One aspect with potentially large explanatory power for both violent entertainment and 'real-life' reality selections is *sensation-seeking*, a personality characteristic (Zuckerman 1979): "An undeniable characteristic of violent imagery is its emotional wallop. It gives people a jolt. Not everyone finds this kind of stimulation pleasant, but some do" (Goldstein 1999, 276). Is it actually *pleasant* for some to watch frightening, brutal scenes? Then the measurement of experienced utility would not be a problem because people would report *positive utility* for the respective scenes? I am not sure. Isn't this sort of entertainment always juggling at the boundary between negative and positive emotion, and wouldn't this imply a fluctuation between negative and positive utility? And what would a person then report in terms of experienced utility when watching a violent movie?[13]

This all seems to point at something that has, to the best of my knowledge, not systematically been integrated into the decision sciences so far: the motivation theory of the *optimal level of arousal* (Yerkes and Dodson 1908). According to this theory, individuals' performance is best with some specific level of arousal, i.e., the performance drops when the arousal falls below or rises above that level; and individuals actively seek out stimuli that *induce* a certain level of arousal they feel is optimal for them, either by reducing the stimulation via certain stimuli or by increasing it. The neglect of this phenomenon in the decision sciences might be viewed as surprising since this theory, also known also as the *Yerkes-Dodson law*, is around for more than a century and empirically supported. However, only recently a good theoretical explanation via the effect of certain stress hormones

---

[12]As already argued in another footnote, above, the reasoning pro 'voluntary' acts is less straightforward with vectorial choices when each reality gets an—at least minimal—amount of consciousness allocated to it.

[13]I am not aware of any studies actually looking at this.

(glucocorticoids) has been postulated (Lupien et al. 2007, 218; see especially Fig. 4 on that page). Thus, it might well be that the reception of this theory in the decision sciences is still ahead. It is important to note that *if* different individuals produced those hormones either to a different extent or reacted to them differently, then different propensities to seek out strange experiences would follow.[14] Given the complexity of the situation, it is quite clear that the optimal level of arousal might imply some *non-linear* impact of certain factors on utility levels on the one hand; on the other hand, it is unclear how such a relationship might exactly look like in terms of experienced utility/disutility over time (and with different individuals).

A related phenomenon, dealt with in the literature on *consumer behavior*, is called *variety seeking* (e.g., Kahn et al. 1986; Kahn and Isen 1993). Straightforward examples for this type of behavior stem from food and music: Let me assume that you have a preferred sort of ice-cream: vanilla. Whilst vanilla ice-cream would—by definition—generate the *highest utility* for you *in general*, you will probably agree that you would sometimes (or quite often) also eat chocolate or strawberry or some other sort of ice-cream. That implies that you voluntarily choose options, e.g., chocolate ice-cream, with lower than maximum utility because there is a *utility from changing back and forth* between vanilla ice-cream and chocolate ice-cream or from buying a cone with the two sorts of ice-cream; or, alternatively (but with the same result), you experience *boredom from not changing* or not mixing experiences, i.e., if the same (maximum utility) behavior is selected over and over. The same holds with music: You will not listen to your favorite song all the time.

Thus, variety seeking is another example where experienced utility alone will not help understanding how individuals form preferences for the realities they select; or experienced utility would have to be defined and measured in a fairly complex way, i.e., context-dependent with eating vanilla after chocolate ice-cream, eating vanilla after vanilla ice-cream, eating vanilla after strawberry ice-cream etc. But practically, there might be a vast amount of possibilities to be considered. Another possibility of looking at the phenomenon of variety seeking, perhaps even the more important one in the multiverse, is putting considerable conscious emphasis on some, non-preferred realities, in parallel to a lot of emphasis on the preferred ones, in *vectorial choice*. Or in other words, you eat vanilla ice-cream this afternoon in one reality, but chocolate ice-cream in another reality. This type of example helps making more sense of the potential dimension of utility interdependencies, one of the determinants of vectorial choice (Tables 8.4 and 8.5 in the last chapter). I am going to consider this and related phenomena in the next section.

Coming back to the literature on films and related entertainment, there is another motive that might play a role more generally, when it comes to explaining the enjoyment, or more neutral: partial selection of the respective reality in vectorial choice, of brutal or odd experiences: the *justice motive*. "Viewers come to have

---

[14]Recent findings (October 22, 2015) reported on the Centre for Studies on Human Stress website, http://www.humanstress.ca, accessed November 11, 2015, seem to demonstrate interesting differences in their biological responses to stress between gay men, lesbians, bisexuals and heterosexuals.

strong feelings and fears regarding protagonists and antagonists and decide in moral terms what fate they deserve. (…) [The] typical storyline of enjoyable entertainment involves the establishment of animosity toward wrongdoers, which makes later violence against them seem justified and hence enjoyable" (Goldstein 1999, 279). It is clear that such motives—reasonable or not—may play a role outside the cinema or video games as well and may drive the selection of certain realities, although the experienced utility in those realities would be quite low or negative outside the *context* of justification of certain behaviors by moral standards or judgment. *Context matters*, so that experienced utility, once more, has to be seen as context dependent and has to be defined and measured in a context-dependent way. The question is whether this concept might then be a good candidate for usage within a multiverse framework. Understanding individuals' vectorial choices would apparently require *the knowledge of all relevant context factors and their exact influence on experienced utility*. This appears to be hard to impossible even in simple situations.

Note that variety seeking and other context dependencies are equally problematic with *decision utility*, however, because the idea of *object-oriented preferences* typically assumed in rational choice is rejected by those phenomena, too (for theoretical possibilities as to how context dependencies might instead be modeled see, e.g., Tversky and Simonson 1993).

Variety Seeking, Sensation Seeking, and Utility Interdependencies between Different Versions of the Individual

In the last chapter and at the beginning of this chapter, utility interdependencies between different versions of the individual have been considered. In the last section, a relationship between utility interdependence and variety seeking has been discussed. Indeed, a fairly complex perspective on utility arises if one assumes that the utilities of different versions of an individual are *not independent* from each other but are somehow connected via invisible and unconscious 'bridges.' If those 'bridges' would require any *physical* connection, however, the idea shouldn't have been part of Tables 8.4 and 8.5 and should not have been dealt with in the current chapter; it should instead have been rejected right away. In quantum mechanics, most would say, there is simply no room for information transfer between different parallel realities.[15] Anyway, this is not what is required here. The connections are supposed to exist between different conscious entities (i.e., versions) of the *same individual* across different realities. This is, I would like to argue, a slightly less problematic idea, also underlying, by the way, the quite essential—for the analysis

[15]Some might argue, however, that at least the interference phenomena at the double slit are associated with somehow interacting, different realities (e.g., Deutsch 1999). I am not sure, however, how Deutsch's conjecture would have to be evaluated within a consequent application of the wave function (without any recurrence to particles).

in several chapters of this book—desire to stay away from very-low-consciousness situations (see Chap. 4).

Assuming that not only information transfer, but also a utility interdependence between different versions of the same individual is possible, there are two pronounced possibilities as to how this interdependence might be 'set up'[16]: (a) utility is *positively correlated* across different versions of an individual, (b) utility across *all* versions of the same individual obeys some *overall criterion* such as variety of experiences, possibly related to the concepts of variety seeking or sensation seeking introduced above; all examples mentioned in connection with Tables 8.4 and 8.5 as well as in the last section of this chapter fall into this category.

(a) The effect of a positive correlation is simple: If many of your parallel versions are unhappy, you are also a bit unhappy.[17] You might be making the right choices, you might be doing everything you can do to be happy, and you partially succeed, but there might still be some, perhaps small, 'background' unhappiness that you can feel. On the other hand, if many of your parallel versions are happy, you might even feel some 'background' happiness if you personally make the wrong decisions. Note that I am repeatedly speculating about the testability of hypothesized relationships between different versions of the individual; the task is all but trivial and surely depends on smart ideas of testing the relationships indirectly via the version of the individual located in *this* joint reality.[18]
(b) Looking at an overall ('across-versions') variability of experiences, one would first of all have to come up with a good reason as to why such a criterion could make sense. So, let me simply assume that *we are here on earth to make experiences*; it would then be plausible to make many of them. Also, one would have to say a word on how large the 'pressure' by such an overall criterion on a singular version might be, given free will of the versions. Let me thus furthermore assume that there is some long-term pressure on the variety of experiences by different versions, but that a singular version is always free to choose otherwise short-term. This would lead to a situation where there is some tendency to make odd, painful and disturbing experiences by some of the versions, but where utility would be hard to understand on the version-level but only taking into account all in a group of versions. This would leave us with a situation that is hard to tackle within any current decision-theoretic framework

[16]Certainly, there are more cases possible, including combinations of those two.

[17]For the sake of simplicity, I am equating here low utility or pain with unhappiness. This is a bit imprecise (see also the last section of this chapter), but not too far from a correct view of things.

[18]Mensky (2010) would probably argue that even a direct contact to other versions of the individual is possible during sleep, meditation or other altered consciousness states, and I could imagine that he is correct, that, e.g., dreams grant an access to parallel realities. However, this is not the controlled access needed within (traditional) scientific research, sufficient for straightforward scientific experiments (at least not given our current state of knowledge). Could, perhaps, MRI or EEG, measured during sleep, grant an access to the information that is needed for such 'inter-version' studies?

but also quite practically; this is the main reason as to why the 'interdependent-utilities' row in Table 8.5 was left blank.[19] But is this all I can say here? I have just (as well as above) mentioned the possibility of, perhaps, empirically (and indirectly) testing those relationships within smart experimental designs in the future. Only such empirical tests would be helpful to 'feed' the respective decision models with information about actual utility interdependencies. This would then be the basis to tackle those interdependencies.

Utility within the Effectuation Framework

In the last chapter, the effectuation approach turned out to be a framework of decision making that could—at least given the current state of theoretical development, but perhaps also principally—be appropriate for usage within the clustered-minds multiverse; it was also mentioned that it might provide a smart way of dealing, in a simplified manner, with the real-options structures often implied by choosing entire realities (despite the possibility of consciousness reallocations that partially demagnify the problem).[20] Within the effectuation approach, a *selection between different effects* is supposed to be made via *satisficing* (e.g., Simon 1956; Tietz 1988). Oftentimes, the focus within effectuation is on objectives such as keeping the potential loss affordable or just 'survival,' e.g., via not entering realities with high consciousness that imply the *option* of bankruptcy (in economic decisions) or personal failure (in private decisions) with certain developments or later problematic decisions (see also Table 8.3). Albeit this is plausible, it should not be overseen that a decision maker has a *reason* to engage in decision making in the first place, and that is to generate utility. Although not in the focus of publications on effectuation, *satisficing* might be applied with respect to the *utility of effects*. Within vectorial choices, *threshold levels of utility* might be taken into account either on the level of each version or on the level of the individual.[21] In the simplest possible view, realities that generate utility above that threshold should be given enlarged subjective emphasis, realities below that threshold should be given a reduced subjective emphasis.

Let me now look at *the* paradigmatic example of effectuation: Somebody is trying to cook a dinner for friends and just looks into his fridge and kitchen cabinet

---

[19]It would also add further complexity to the question of responsibility, a problem that could partially be solved via long-term, self-forming actions in Chap. 7.

[20]As has already been mentioned, later reallocations of consciousness might not be thought of as completely removing a real-options problem. E.g., if all high-consciousness versions of an individual have entered realities (via previous decisions) where a certain outcome is not possible, anymore, reallocations of consciousness would have to be implausibly substantial (in too 'short a period') to be implemented, to still achieve that outcome with high-conscious versions.

[21]It is unclear how this is organized; intuitively plausible would be a combination of both.

to see what is in there to be used for that dinner.[22] Accidentally, he might now find all the ingredients needed for a *Spaghetti Carbonara* such as eggs, bacon, cream, spaghetti etc., but might simply *not like* the taste of Spaghetti Carbonara himself or associates with it a very low utility because he is just trying to keep a diet. In this case, using effectuation as a decisional framework, *avoiding* to cook the Spaghetti Carbonara requires the usage of a satisficing rule with respect to utility. Some minimum utility must be provided within a certain reality to be emphasized by consciousness.

Whereas it became a bit clearer by now how utility considerations could be integrated into the effectuation framework (and how effectuation could be used within the framework of the clustered-minds multiverse, for that matter; see Chap. 8), are there any good reasons to think that experienced utility *works* within the effectuation approach? (The already negative 'balance' I have reached above with respect to decisional utility is even aggravated by the fact that the entire framework of effectuation is based on bounded rationality, and decisional utility assumes perfect rationality; all other criticisms—including the other problems with assuming perfect rationality—prevail.)[23] Are there any new aspects to be detected here that have not been part of the general discussion above? One might argue that problems in connection with experienced utility are alleviated by the fact that utility would *only* have to be used within a *satisficing* criterion. With respect to a minimum threshold, the information and precision requirements are probably smaller than with a continuous dimension that is supposed to be used within maximization. However, the main problem that was detected above, individuals' choices sometimes aiming at odd and painful experiences, is not any more plausible when using experienced utility within the effectuation framework than in any other context; one might have to look into utility interdependencies, as has already been argued, but this is a thought independent of whether effectuation is used as a framework or not, and it is as difficult to implement within effectuation than within any other approach.

Last but not least, the real-options structure of many vectorial choices is just more explicit within decision utility with its ingrained idea of optimality than it is within experienced utility; but the lower transparency in this regard is nothing that could be used as an argument in favor of the experienced utility approach. The reason as to why effectuation is better in dealing with real-options structures—the stepwise, low-commitment approach taken, trying to keep positive options open and cutting on the negative options, possibly via 'self-binding' actions—turns out to be a feature of the framework, mostly independent of the utility framework finally opted in favor of.

---

[22]As already mentioned in Chap. 8 in one of the footnotes on effectuation, the causation approach for the 'dinner problem' would be to select an optimal meal from a cookbook and then buy all the ingredients at the supermarket.

[23]As has been said, already, the main problem is not the rationality assumption underlying normative decision theory and that effectuation is rather a 'mixture' of normative and descriptive approaches, but the inferred preferences assumption of decisional utility.

Final Discussion

This chapter has been concerned with the question as to what existing utility concept (if any) is appropriate for usage within vectorial decision making, based on the clustered-minds multiverse. The final discussion section now aims at condensing the answer so that the state of affairs as well as the requirements for future research become transparent. This requires dealing with multidimensional utility concepts as well as happiness research, so that the final discussion section will be somewhat longer than usual.

Regarding the usage of decision utility, four problematic issues could be detected. Three of them arose from the assumption of *perfect rationality* together with the idea that *preferences* (and therefore utility) can be *inferred* from choices. Already outside the multiverse (i.e., within standard decision theory), the fact that people are not perfectly rational makes it impossible, as required, to infer preferences from choices. However, using the knowledge from behavioral decision making (see above), one could try in principle to *correct* the preferences revealed in choices to infer the *actual* preferences of the respective individuals. Those corrected preferences could then be used to support individuals in their decision making (normative goal), they could also be used to better predict people's decisions (descriptive goal). This potential solution, however, suffers from the *multiplicity* (status-quo effects, loss aversion, context effects and so on) and *complexity* (interplay of the factors in the last parenthesis, for instance) of deviations from rationality in real decisions. The second issue is multiverse-specific. It arises from the fact that decisions are made differently by different versions of the individual, the main characteristic of vectorial choice. And how could one possibly infer preferences from the decision of one version of the individual if there is no perfect coupling between preferences and the observed decision of that individual? This problem cannot easily be solved in the multiverse, or at least only long-term, i.e., with multiple decisions (see, for a related issue, Chap. 7). The third issue is that perfect rationality implicitly assumes non-myopic behavior and therefore, given the actual branching structure implied by multiverse choices, asks, eventually, for a real-options modelling (Dixit 1992; Dixit and Pyndick 1994); this holds, as has been discussed a few times (partially within footnotes), despite the fact of possible corrections by reallocations of consciousness, if those corrections are assumed to take place in a realistic range. Given the high complexity of vectorial choices to start with (revisit, again, the already simplified structure of determinants in Tables 8.4 and 8.5), it is hard to imagine how to explicitly integrate this into a quantum decision making concept, actually with the time horizon of a lifetime. Effectuation with its pragmatic structure offers possible ways to deal with this issue in simplified ways—and independent of the utility concept finally assumed.

The fourth issue with decision utility, partially related to the third, however, turned out to be of different nature. It arose from the fact that the 'high-frequency'-type experience of specific realities would require a *high-frequency measurement of choices*. A definition, explanation and formal modelling of split-second decisions would require a detailed understanding as to what makes people deciding certain things in certain ways that quickly—and what it actually *means* to make such a decision. Normally, we perceive only the 'big' decisions as such: marrying the right person, selecting the right job, buying the right house etc. But what the split-seconds perspective on choices, inspired by the clustered-minds multiverse, would tell us is that the decision to leave the house for lunch now or rather in five minutes, leads to a residence in a different universe. And that might be more important (under certain circumstances) than buying one house or another.

Whereas the three problems reported in connection with the perfect rationality and revealed preferences assumptions are *excluding* the usage of decision utility as a concept to be used within quantum decision making, the fourth problem rather poses an interesting and important challenge to be addressed in future research on quantum decision making. It is an issue that simply *has* to be addressed at some point, no matter what utility concept will finally be used.

Given the difficult situation with respect to decision utility, one might have expected experienced utility to be the solution. Three issues remained open here after the intensive discussion above. The first issue is voluntary choices of 'negative' realities by individuals. It is a major problem with experienced utility because it is completely unclear what it implies to voluntarily choose negative realities in terms of experienced utility: a negative experienced utility that is preferred over a positive one; or a positive experienced utility from negative experiences? It appears as if the basic premise by Bentham that the "goodness or badness of experience is the pleasure or pain arising from the experience" (Read 2007, 46) were at stake. A possible solution, to be explored in future research, lies in the appropriate usage of sensation-seeking tendencies, optimal level of arousal motives, and variety-seeking motives (first concentrating, for simplicity, on a singular version of an individual and abstracting from potential interdependencies with other versions) in the determination (correction?) of experienced utility. (Another possible solution might be the integration of multidimensional utility concepts; see below.)

The second issue that has been detected is the fact that remembered utility often deviates dramatically from experienced utility and that it is unclear which of the two concepts would be underlying individuals' vectorial choices; this problem partially appears, however, in a singular reality already. One might argue that this is primarily a matter of empirical research that perhaps relates the two concepts to something overarching such as satisfaction with life or even happiness and measures the strength of relationships between the respective criterion and the two; and perhaps this is the way to go. But there are also two problems arising with this idea. One is the fact that Kahneman's 'translation' of experienced utility into 'objective happiness' (e.g., Kahneman 2000) has been criticized so that it is inherently unclear whether there is a direct relationship between the two (Alexandrowa 2005; see the general discussion below). The other is that the problem of allocating large amounts of consciousness to negative realities in vectorial choices is returning through the backdoor: Might a negative experience actually lead to a positive utility and then to happiness? Or will the negative experience be leading to a negative utility and then to unhappiness? Or is all this context-dependent or, again, 'mixture-dependent' in the sense that the combination of realities matters? Finally, and quite normatively: What does it really *mean* for an individual to be unhappy with a situation that has been *chosen* by that individual? (Note that this has nothing to do with lack of knowledge before and surprises after a singular decision. I would rather look at long-term, repeated choices where an individual voluntarily stays in a situation that makes him unhappy.)

The third issue arose implicitly, mainly from the discussion of decision utility. Experienced utility kind of 'hides' the lack of a necessary *future orientation* of decision makers or, specifically, the real-options structure of many choices, already in a singular universe, but even more so in the multiverse. Unless explicitly modelled in a novel, fairly complex framework, experienced utility implicitly assumes myopia of the decision maker; perhaps, but this is a fairly notional statement, a clever account of *anticipated experienced utility*, integrated in an appropriate mathematical framework, could be a way out.[24]

I have only briefly mentioned multidimensional utility concepts so far. What usage could be made of multidimensional utility concepts within quantum decision making? (For the sake of reducing complexity, I am disregarding, in the following discussion as well as in the discussion of Kahneman's 'objective happiness' concept following it, the problem of real options, the restrictions to implement only moderate allocations of consciousness and to avoid very-low-consciousness situations as well as the possibility of reallocations of consciousness.) Could any of those concepts be helpful in modifying or refining some of the existing utility concepts? Let me explore this thought a bit. There are many multidimensional theories existing—not to be confused with the vectorial decision making concept proposed in this monograph that is concerned with the multiplicity of realities. A plethora of approaches on multicriteria decision making is used in operations research; the questions which apartment to rent or which house to buy or how to transport cargo are typical for this approach; they might, however, be seen as less relevant for a conceptual discussion (see, e.g., Triantaphyllou 2000, for an overview).

The theory of reasoned action (Fischbein and Aizen 1975) is often not read as a utility concept; but I view it as one, albeit not of the microeconomic type—and without the normative appeal achieved via explicit, underlying rationality axioms. Another multidimensional utility concept stems from the microeconomic theory of consumer demand (Lancaster 1966). According to Lancaster (1966), people do not buy goods but vectors consisting of, say, vitamins, carbs, fats, proteins etc. Consequently, the utility function is also multidimensional.[25] What can be used here is Lancaster's (1966) idea (as well as Fischbein and Aizen's 1975) that there might be tradeoffs that individuals make between attributes when choosing between different realities. I would like to argue that formally, Fischbein and Aizen's (1975)

[24]The idea of *anticipated experienced utility* might be less absurd than it sounds at first sight. In the decision sciences, anticipations like this are common. A fairly well-known example is *anticipated regret theory* (Loomes and Sugdon 1982).

[25]I am abstracting here from the fact that Lancaster (1966) is analyzing consumer activities incorporating own actions (household production) by the consumer.

concept fits better with the vectorial decision making approach, because of its linearity. According to Fischbein and Aizen, alternatives are evaluated via their adequacy to satisfy an individual's need (often elicited with simple rating scales), measured on the level of multiple relevant attributes, multiplied with an attribute-specific weight, indicating the importance of that attribute for the decision maker. The sum of all importance-weighted, adequacy-rated attributes is the criterion with the help of which alternatives are compared.

Specifically, combining vectorial decision making and—already weighted—attributes,[26] according to the Fischbein and Aizen (1975) approach, would generate a two-dimensional matrix, potentially—with some additional premises—depicting the problem of decisional conflicts and vectorial choice at once; individuals could then optimize the utility generated by the entire matrix, if the underlying functions and restrictions, e.g., specifying utility interdependencies etc., would be known. Nothing should be in the way, also, of interpreting the attributes as describing *experienced utility components*, I suppose. Oftentimes, restrictions and utility dependencies might just be brought in in the form of an additional vector: E.g., the size of a decisional conflict (one of the determinants in Tables 8.4 and 8.5), given by large attribute weights attached to rather inferior and, simultaneously superior attributes with *one* reality could be integrated into the vectorial choice by such an additional vector. This all constitutes, however, a serious research agenda, far beyond what could be accomplished in this monograph.

I have already stated that the existence of a higher-order criterion such as happiness would be helpful in deciding upon the feasibility of different utility concepts (or perhaps even replacing them, if the definition were fine grain enough to be used at the level of choices). Unfortunately, defining and measuring happiness is not simple, either. I would like to concentrate here on briefly defining and criticizing Kahneman's 'objective happiness' concept because this will be quite revealing in this regard. According to Kahneman (2000, 681): "(…) objective happiness is a moment-based concept, which is operationalized exclusively by measures of the affective state of individuals at particular moments in time." Or in other words, *Kahneman's concept of happiness builds up upon experienced utility directly*. And this is exactly the problem: "Kahneman's methodology precludes incorporation of relevant pieces of information that can become available to the subject only retrospectively" (Alexandrowa 2005, 301). In a simplified view, it is "(…) crucial that happiness at the level of experience is taken to be only one component of (…) [subjective well-being], to be supplemented with an active reflective endorsement sometimes called life satisfaction. (…) [Subjective well-being] is thus taken to encompass happiness both as experience and as an attitude" (Alexandrowa 2005, 302; see for a similar reasoning Diener and Lucas 1999).

So, if happiness is understood as subjective well-being (as is often supposed), it falls into two categories. One is quite similar to experienced utility, one *requires*

[26]This would generate singular (attribute-specific), importance-weighted adequacy measures, contributing to overall 'utility.'

reflection upon the past. If one were to compare the relationship between happiness and experienced utility on the one hand as well as remembered utility on the other hand (as has been discussed above), what would happen, and how meaningful would this result be? Sure enough, experienced utility would have a positive correlation with happiness; in fact, the measurement of this component within happiness might look quite similar or even identical to the one used to measure experienced utility (Alexandrowa 2005). Remembered utility would exhibit some (possibly small) correlation with happiness. The reason is nothing substantial but merely the fact that the measurement (and construct) of remembered utility and the reflective part in happiness are very different. Whereas the latter would perhaps be measured via the Satisfaction with Life Scale (Diener et al. 1985), remembered utility would be measured as described within the colonoscopy example above: The colonoscopy patient would be asked for his experience after the colonoscopy is finished. The respondent filling out the Satisfaction with Life Scale would be asked questions such as "In most ways my life is close to ideal." If those two are different, is it really necessary to have a reflective component in the construct of happiness such as questions about life in general? Yes, it is. And Alexandrowa (2005) has a couple of nice examples for this demonstrating how important the subjective, reflective component in the evaluation of one's own happiness is.

Another criticism by Alexandrowa (2005) is as important as the one just discussed, given the question I have raised earlier on the—perhaps—fundamentally flawed idea that in experienced utility 'negative' events might generate 'negative utility.' Specifically, Alexandrowa (2005, 306) criticizes the assumed uni-dimensionality of the good-or-bad dimension and discusses examples such as running a marathon or reading a tragedy clearly rejecting this uni-dimensionality from her perspective. Let me now bring in, again, the idea of tradeoffs between attributes (Fischbein and Aizen 1975). Perhaps, 'negative' and 'positive' aspects of an experience might get different weights, and the 'negative' aspects get a relatively small (but positive) weight, like the salt in the soup. This might hold on the level of a singular reality as well as across realities. And this also has to be explored in future research.

To conclude, there is currently no answer available to the question as to what is the appropriate utility concept to be used within quantum decision making in the multiverse. It is clear that none of the existing concepts fully suffices in this regard. Certain aspects of experienced utility and multidimensional utility might be used and appropriately combined, forming a matrix, with vectorial choice. But, as has already been mentioned, this is rather the start of a research endeavor than some situation close to the end of it; as is the case in some other chapters of this book, I was able to identify a number of *specific* research opportunities. Note that the complexity of this research agenda would eventually even be higher than what could have been conjectured based on the last two discussions on multidimensional utility and happiness. The reason is that I explicitly left out considerations with respect to real options, moderate allocations of consciousness and avoidance of very-low-consciousness situations as well as reallocations of consciousness. Combining those frameworks and restrictions with what has been said is not manageable at this point, or perhaps only within a specifically dedicated monograph.

Chapter 10
Games and Markets

Strategic Choices and the Multiverse

In the last two chaps. (chaps. 8 and 9), decision making has been analyzed in *non-strategic* situations.[1] However, in the multiverse other players might enter the picture already in those setups because of the need of avoiding 'very-low-consciousness' situations. Also, other players may implicitly become part of the 'other factors' (see chap. 8). A large step beyond that type of involvement of other individuals are *strategic decisions* where the decisions of (intelligent) counterparts are *explicitly* considered, in situations with and without closeness to the fringe of a consciousness cluster. Since the avoidance of very-low-consciousness situations plays a major part in the considerations of this chapter, too, already leading to a high complexity, reallocations of consciousness as well as the requirement to choose moderate allocations of consciousness are not considered here. Both factors, are, anyway, too vague at this point without further qualification in future research, for a usage within a game-theoretic discussion.

Strategic situations are normally analyzed in game theory and in economics (two related and largely overlapping fields), and such analysis typically has a high degree of mathematical development. The results of the analyses are called *equilibria*; equilibria can be interpreted as a normative *prescription* of behavior of individuals in the respective situations, but they are often seen—sometimes without further reflection—as behavioral *predictions*, too (this assumes a high degree of strategic rationality of the players). What happens if such strategic/market situations are analyzed within the framework of the clustered-minds multiverse? This question is in the middle of this chapter. Note that the complexity of strategic situations, involving choices of players *and* their counterparts, is *much* larger than in regular game theory in a singular reality. All strategy combinations in a game are played in parallel in different realities, perhaps

[1]Decision scientists would often call those types of decision situations 'games against nature,' but I find this phrasing unpleasant and inappropriate, given that mankind—in light of climate change etc.—might actually be seen as playing games against nature in a quite literal sense.

© Springer Nature Switzerland AG 2018

C. D. Schade, *Free Will and Consciousness in the Multiverse*,
https://doi.org/10.1007/978-3-030-03583-9_10

with low consciousness being involved from one or more sides. An interesting question is what the equilibria suggested in game theory have to say in this respect.

A large part of the chapter focuses on the re-analysis of an exciting experiment within the paradigm of *simultaneous market entry*, repeatedly played, but not a dynamic game, because it is especially well suited to connect strategic/economic thinking with a multiverse perspective.[2] This type of decision situation has been suggested by Selten and Güth (1982), it can *formally* be analyzed within the framework of game theory, and it has important consequences for economics, more generally. I am looking at a version of the game analyzed in an experiment by Rapoport (1995). Rapoport has been, in turn, inspired by an earlier market-entry experiment carried out by Kahneman (1988) and builds up directly on him. Kahneman was very surprised by the set of findings he observed in a classroom experiment and wrote: "To a psychologist, it looks like magic" (Kahneman 1988, 12). Since Rapoport's results turned out to be about as 'magic' as Kahneman's, I wondered what those findings might imply when they are analyzed within the framework of the clustered-minds multiverse.[3] This section has three subsections. In the first subsection, I will explain Rapoport's game and experimental findings and show why they might be called 'magic.' In the second subsection, I will then translate the simultaneous market entry situation into the framework of the clustered-minds multiverse. In the third subsection, I will discuss whether the challenging findings in those market entry games help us to better understand the workings of the multiverse in turn, whether something more generally might be learned from them. Another long section is speculating on relationships between economic games and the multiverse beyond Rapoport's game. In the first subsection, another version of market entry is analyzed within the framework of the multiverse, the skill-based game studied by Camerer and Lovallo (1999). In the second subsection, experiments comparing behavior against computers with games against humans will be analyzed through the lens of the multiverse. The chapter ends with a section containing some concluding remarks. This chapter might be seen as a first step towards *strategic quantum decision making* or *quantum game theory* within a multiverse framework.

A Re-Analysis of Simultaneous Market Entry

Playing Rapoport's (1995) Game

Imagine the following situation. You and fifteen other individuals are sitting in a classroom and participate in an experiment. The experimenter stands in front. The

[2]This type of game also allows abstracting from a real-options structure that would otherwise have to be taken into account or the equally complex issue of dynamic equilibria within the multiverse.

[3]Others have tried to clarify what is going on in those experiments using 'classical' means (including different experiments, other statistics etc.), however, with mixed success. It is beyond the scope of this book to go into those papers and results, here.

experiment results in monetary payoffs. The *monetary payoff* that you (as well as the fifteen other players) achieve in this experiment is *contingent on performance*. The cash amounts that you and the others will earn will depend on your own choice and the choices of the others. The game will be played over multiple trials (i.e., 20), and you are handed out a response sheet where you have to *anonymously* specify your choice in each trial.

As already mentioned, the experiment is about a certain version of a *simultaneous market entry game* (Rapoport 1995). The situation played in each trial can be described as follows. Each individual (including you) is supposed to decide whether or not to enter a 'market.' If you stay out of the market, you get \$1.00—with certainty. If you enter the market, your payoff—the amount of money you will receive in cash—will precisely depend on two things:[4]

(1) It depends on the so-called 'capacity' of the market, c. The capacity varies between 3 and 12; it is randomly chosen by the experimenter and written on a board.
(2) It depends on the number of players, e, including you, that *actually decide* to enter the market.

Specifically, a player's payoff when entering the market is determined by subtracting e from c and multiplying the result with 2. The resulting number will be interpreted as a dollar amount and \$1 added to it. This is the payout. E.g., if the market capacity is 3, but 5 players decide to enter the market: $3-5=-2$. Multiplying this by 2 and adding one, the payoff for each of the five players who entered is $-\$3.00$. The 11 players that stayed out of the market will each end up with \$1.00.

Thus, if there is a large excess entry, all individuals that entered lose a lot of money. If fewer people enter than the respective market capacity, everyone who entered earns money in excess of the \$1.00 (he would receive when staying out). The largest amount of money a player can make in a given situation occurs whenever he is the *sole* player that entered.

What is the behavioral prediction for that game? The situation is tricky, even if using the help of normative game theory. Applying the notion of a *Nash equilibrium*: players selecting *mutually best responses*, where no player has an incentive to *solely* deviate from that solution (von Neumann and Morgenstern 1947; Nash 1951), the simultaneous market entry game is a *coordination game* where *multiple* such Nash equilibria exist in *pure strategies*.[5] Casually speaking, many combinations between entry/non-entry decisions are possible that lead to an equilibrium. Specifically, there are two situations that qualify as an equilibrium condition, and

[4]In fact, due to the large possible payoffs, respondents were paid out based on the payoff achieved in one of the rounds in Rapoport (1995), only. That round was, however, selected randomly after all experimental trials had been completed.

[5]The two pure strategies in this situation are 'enter' and 'do not enter.' Mixed strategies would in contrast specify a probability distribution between the two (one of the two strategies would be 'drawn' rather than selected)—see below.

each of them can be reached by numerous player combinations in turn: all players' strategy combinations where the resulting number of entrants is *equal* to the market capacity and all players' strategy combinations where $e = c-1$.[6] *Who* of the sixteen players are those that are in the market and those that stay out is *irrelevant* from the perspective of game theory, contributing to the multiplicity of equilibria. But it is relevant for whether one of the many equilibria will be reached in actual play: Coming back to the experimental instructions, players *anonymously* choose whether or not to enter. Any possibility to coordinate their entries via communication does not exist. So, it is completely unclear which combination between entry- and non-entry-decisions will be played, i.e., *who* is staying out and *who* is entering. Is there any other possibility that normative game theory offers for this case? Yes, indeed, it is the *Nash equilibrium in mixed strategies*. With this solution, players would have to determine a certain *entry probability* that would directly depend only on the number of potential entrants (i.e., 16) and the respective market capacity. Their entry decisions would then be resulting from an *internal random draw* applying this probability.[7]

What did Rapoport (1995) find in his experiment?[8] "(…) Individual differences are (…) substantial. (…) Subject 3 never entered the market, Subject 9 entered it only once, as did Subject 12, whereas Subject 11 entered it on 15 out of 20 trials. (…) As noted by Kahneman, *e* varies with *c*; indeed, the product-moment correlation between *c* and *e* (based on all 20 trials) is positive and highly significant ($r = 0.83$, $p < .01$)" (120). An important feature of the experiment was that "Values of *c* were sampled without replacement from the set {3, 4, …, 12} and announced one at a time on trials 1 through 10 (Block 1). The same sampling procedure was repeated on trials 11 through 20 (Block 2)." (ibid.) Since each market capacity was selected twice, Rapoport was able to look at the "within-subject variability in strategy" (ibid). Specifically, he was able to check whether individuals played a *pure strategy*. This would be indicated by individuals *repeating* their strategy when the same market capacity occurred for the second time (the play of a mixed strategy would have 'destroyed' the repetition pattern in most cases). And this turned out to be true in an overwhelming number of cases. Out of the 160 resulting 'pairs' (16 respondents playing each capacity for ten times; but each capacity was presented twice, hence a 'pair'), 141 were repetitions, only in 19 cases the strategy was not repeated.

---

[6]The latter case is special because the 'next' player is indifferent between entering and not entering.

[7]An explanation as to how to calculate the specific probability is beyond the scope of this book. This also applies to the discussion as to how those random draws could exactly be understood/ interpreted.

[8]Indeed, Rapoport repeated this experiment, with some weeks in between the first and second and the second and third 'block,' with the same subjects (participants in a Ph.D. class). So, in total, he collected data from 60 trials (with only very few no shows in the second and third block). The data from the three blocks differed with respect to the overall entry, compared to capacity. But they did not differ with respect to what I am interested in, here. Therefore, I concentrate on the analysis of the data in the first block, i.e., restrict the analysis to the first 20 trials.

Those results are indeed 'magic,' or at least challenging from the perspective of game theory and, more generally, in a singular universe. As already noted, due to the multiplicity of equilibria in the simultaneous market entry game, pure strategy equilibria simply cannot be a feasible solution of that game if communication between players is impossible. Nevertheless, individuals *do* play pure strategies, and they *do* coordinate somehow as indicated by c and e being closely related. How are those findings to be explained?

Simultaneous Market Entry as a Game Played in the Multiverse

In the multiverse, all possible outcomes of a simultaneous market entry game coexist in different, parallel realities. However, some of those realities are characterized by very-low-consciousness situations, others contain a lot of consciousness. And not only those combinations of players' versions playing in a way that leads to an equilibrium exist, but also each combination not leading to an equilibrium. Let me look at a specific situation in the above game: A market capacity of 3 has been announced in the Rapoport (1995) scenario and the decision you have to make is whether or not to enter the market. How would your considerations look like in the multiverse?

Let me assume that you want to maximize your monetary outcome from that game, e.g., because you plan to spend it on a nice dinner with your significant other. Then why don't you just allocate a lot of consciousness to this reality, i.e., the reality where you are the only entrant in the market and all others stay out? In this case, $c-e = 2$. Multiplied by 2, plus 1, you will end up getting \$5.00. Not much for a dinner, not even for one person, but that was the maximum you could achieve, here. Clearly, however, not all versions of yours are experiencing this reality. But, in the multiverse there are no principle limits to a (partial) shift of emphasis towards this reality since it principally exists, or are there? (As said in the beginning of the chapter, the restriction of moderate allocations and the possibility of reallocations of consciousness are not taken into account in the analysis presented in this chapter.)

Note that I have postulated a fairly 'hard' limit to 'social interaction' in the course of the book, for the sake of preventing ontologically irritating scenarios: *avoidance of very-low-consciousness situations* (see chap. 4). Is this a problem here? Let me go into some more detail. A very-low-consciousness situation occurs whenever you (a version of yours with a high degree of consciousness) are surrounded by close-to-zombie entities. This situation in turn arises here if consciousness is almost entirely allocated away from those versions of the other players that end up in this reality. Now let me suppose the plausible case of players that *all* want to maximize their payoff and that accept, as a secondary reality, only a situation where they share the market with *one* more player; let me finally assume that *you* have the same preference as all the other players. In this secondary reality: $c-e = 1$. Multiplied by two, plus one, each version entering the market would end up with a payoff of \$3.00. This is still better than staying out of the market (that would lead to \$1, only). More precisely, all players put primary emphasis on a situation where they enter the market and where they are the only player in the market, and they still put a lot of emphasis on the reality where they enter the market and share the market with one more player. And they allocate away consciousness (as much as possible) from realities where they share the market with more players *or* (and this is more important, here!) where they stay out of the market. The resulting question now is whether there is any reality where your idea of maximization (or at least your 'compromise') actually works out without violating the very-low-consciousness restriction. The answer unfortunately is *no*. Since no player puts more than minimal emphasis on staying out of the market, situations where you are alone in the market or share the market with one more player(s) would clearly be situations with very-low-consciousness versions of the others.

What happens if the individuals are a little more flexible and still consider the case of sharing the market with *two* others as a preferred outcome? The important change would be that the latter case would make them indifferent with staying out of the market (if they only cared for money what I assume here for simplicity, because in both cases the payoff would be one), and one version of each individual will then do so with larger than negligible degree of consciousness. Consequently, positive payoffs happen in some realities with non-negligible probability and without running into a very-low-consciousness problem. But note that we need a huge number of versions of individuals with more than minimal consciousness, here, and that we are also 'producing' a large number of subjective worlds with 'enough' consciousness allocated to them.

Even though I am not able to come up with a specific *upper limit* for the number of versions of an individual *with more than minimal consciousness*, here, it is clear that there must be *some* limit to 'creating' more and more non-negligible-consciousness versions residing in different outcome-cases in the simultaneous market entry game without 'diluting' consciousness too much (note that in formula (1), describing vectorial choices, the total amount of consciousness is assumed to be given for any next decision, implemented by some version of the individual, resulting in several new versions). An interesting theoretical question is whether one can make any comparative statements such as a statement about *differences* that might result when one moves from a smaller to a larger market capacity in this regard. Is the situation more or less 'critical' with respect to the number of such *required versions* if one keeps the idea of individuals maximizing payoffs? In order to be able to analyze this more general question, I first have to redefine the restriction by moving from the graphic assumption of a maximum number of individuals a player is willing to share the market with (i.e., 2) to the assumption of a *minimum payoff*. This assumes *satisficing* or some aspiration level (the usage of satisficing was also suggested within the framework of the effectuation principle in chaps. 8 and 9). To stay comparable, I will assume, as was implicitly assumed in the second version of the prior example with $c = 3$, a minimum payoff that equals the payoff from non-entry, \$1.00. So, in terms of versions, individuals are willing to allocate a non-negligible amount of consciousness to versions that are at least earning $1.00, but will allocate the largest amount of consciousness to a reality that maximizes their payoffs, i.e., $e = 1$, when they are the only player entering the market.[9]

Let me now assume that the experimenter has announced a much larger market capacity, i.e., 10. In this case, individuals requiring a minimum payoff of $1.00, but trying to maximize payoffs, are willing to allocate consciousness to realities where they are the sole player in the market, where they share the market with one (of the 15) players, with two of the fifteen players, with three of the fifteen players and so on, up to nine other players. All those players can arbitrarily be drawn from the set of 15 players. This leads to an enormous flexibility in terms of acceptable solutions on the one hand, but also to a much larger number of required versions of each individual than in the $c = 3$ case.[10]

As long as the case of staying out of the market is included into the preferred realities, individuals will allocate a non-negligible amount of consciousness to versions staying out of the market that earn only $1. But other versions of theirs (in turn meeting versions of other individuals that stayed out of the market) are able to maximize payoff in a different reality. As has already been mentioned, in the case with $c = 10$ this requires a *stupendous* number of versions of individuals allocated to different realities (i.e., outcomes of the game). This would imply a pronounced 'dilution' of consciousness together with many realities exhibiting very-low-consciousness problems. The only solution would be *consciousness clustering* around certain realities.[11] The premise that consciousness clusters is implied within the definition of the clustered-minds multiverse, already (see chapter 4). But practically, given the framework of the simultaneous market entry game, how could that work?

---

[9]It is a bit unusual in the decision sciences (including effectuation) to have an aspiration level (e.g., a minimum payoff) *and* a maximization criterion applied to the same output dimension. However, in the multiverse such an assumption appears reasonable since different realities are to be 'covered' by consciousness.

[10]I do not need to prove this formally, here, a simple 'verbal proof' clearly suffices. Indeed, the cases of one player being *alone* as well as sharing this market with *one* or *two* other players pertains to both $c = 3$ and $c = 10$. All additional cases such as sharing the market with 3, 4, 5 etc. other players only exist with $c = 10$.

[11]Staying in one reality cluster supposes that the memories of the participating versions are consistent. Violations of this consistency ("You remember us having had an entry of 5 with that capacity, and you tell me that you were one of the entrants; but I remember an entry by only one person, and that was me") are a problem. (At least we do not normally experience them.) Although I excluded consciousness reallocations from the considerations in this chapter, Box 6.3 with its considerations on memory and 'quantum brainwash' are relevant, here.

Clustering within the Clustered-Minds Multiverse: Can Anything More Generally be Learned from the Rapoport (1995) Experiment?

Rapoport (1995) reports on a set of surprising experimental results and, as has already been stated, confirms Kahneman's coordination 'magic.' What is the status of those findings within the framework of the clustered-minds multiverse? First of all, it is *his* findings, so the results occur within Rapoport's reality, partially based on his expectations, and they are then communicated to us.[12] This aspect will further be explored in chap. 12 concerning itself with an interpretation of the Rosenthal effect within the framework of the multiverse. Second, they indicate not only *one* manifestation of reality in the multiverse, but instead the manifestation of several realities because of the numerous replications he ran. Therefore, we may want to see Rapoport's experimental findings as *scientifically relevant* (as any other experiment with a sufficient number of repeated trials), i.e., clearly beyond just a 'snapshot' of one reality out of many possible in the multiverse. Perhaps, we may want to see the experimental coordination outcomes as a realization of experiences within one reality cluster. Third, is the *quantum game theory* perspective, i.e., vectorial choices in a game context, helpful in understanding the 'magic,' the surprising degree of coordination, in Rapoport's (1995) game? How would clustering of consciousness enter the picture? Remember that players hardly exhibited over- or under-entry on the group level in the vast majority of repetitions, although most participants *appeared* to use pure strategies.

Why did I put 'appear' in italics in the previous sentence? The reason is that a *quantum superposition* might be interpreted as something related to a *mixed strategy*, albeit a special one. In the clustered-minds multiverse, individuals might rather distribute consciousness *across versions* of theirs than implementing a random selection of strategies *across trials* with one version (revisit Fig. 4.1, for a deeper clarification of this thought). The question remains as to how the allocation of consciousness to different realities, how the vectorial choice adapts to the market capacity and leads to an overall entry close to that capacity. Experimental participants might allocate consciousness in a way that highly conscious versions of participating individuals cluster within certain realities; their combinations lead to *reasonable outcomes*.

Experimentally, there is a player type that is somewhat 'aggressive' in most trials, tends to 'go ahead' and generates profit mainly from market entry. Another type does not enter in most trials and instead prefers the small income from the fixed payment ($1.00) when staying out of the market. A third, larger group, somehow adjusts the decision on whether or not to enter the market to the announced market

[12]In fact, the situation is even more complex. There are different versions of Rapoport getting different measurement outcomes, and we (I as the author, you as the reader) share a reality with a certain version of Rapoport that generated this type of results. Given the complexity of this thought, I opted for using a simplified version of reality in the text.

capacity.[13] The results by Rapoport (1995) indicate—when interpreted from the perspective of the clustered-minds multiverse—that versions of individuals indeed tend to cluster in *meaningful* ways. This leads to a reduced problem of consciousness 'dilution' because larger amounts of consciousness are allocated to smaller numbers of realities and to a reduced problem of very low consciousness. Not that the number of low-consciousness realities would be reduced (the opposite is the case); instead, the number of conflicts (where low-consciousness situations become a restriction) due to versions of individuals wanting to play those realities with substantial consciousness is reduced. There are simply less realities that people 'want' to play. In the case of the market entry game, the clustering is about different profiles of individuals in terms of 'aggressiveness' or 'risk taking,' in other games (and outside experiments) clustering might be organized through different characteristics, also through multiple characteristics. In any case, quantum game theory, based on the clustered-minds multiverse, tentatively suggests a different 'story' with respect to the coordination 'magic' observed in simultaneous market entry. I am afraid, however, that this 'story' only partially qualifies as an *explanation* of the phenomenon. How exactly the clustering into smaller numbers of realities works must be left open, here. In any case, the observed 'magic' in experiments on simultaneous market entry seems to *suggest some* meaningful clustering—beyond just avoiding very-low-consciousness situations. And this inspires future research both in experimental game theory as well as with respect to a better understanding of the workings of the clustered-minds multiverse.

Lessons beyond Rapoport's Game

Market Entry Once More: The Case of 'Overconfidence'

Let me first stay with market entry situations, but let me now look at the game that was analyzed by Camerer and Lovallo (1999). In their study, experimental subjects also simultaneously entered an experimental market, and their success from entry also depended on the number of other players that entered the market (as in Rapoport 1995), but their experiments were different from the Rapoport (1995) type in a few important points:

> Payoff's depend on a subject's rank (relative to other entrants); ranks depend on either a chance device, or on a subject's skill; subjects in some experiments are told in advance that the experiment depends on skill (and hence, more skilled subjects presumably self-select into the experiment); and subjects forecast the number of entrants in each period. (Camerer and Lovallo, 1999, 308)

Their results are often seen as one of the fundaments of the *overconfidence* phenomenon in economics. Specifically, whenever the skill of individuals played a

[13]In a way, Rapoport (1995) is speculating along similar lines, but he struggles with the fact that this explanation is not convincing in a singular reality.

role, they significantly over-entered the market. And over-entry increased once more when they had *self-selected* into the market. Since everyone in the latter treatments knew that everyone was recruited with the announcement that a certain skill mattered, everyone should have known that the other players opted in favor of participation in the experiment because they thought to *possess that skill.* Or in other words, everyone should have known that competition will be hard. Entering nevertheless, individuals faced a fierce competition. Camerer and Lovallo (1999) coined the term '*reference-group-neglect*' for this phenomenon. It is important to note that in the self-selected, skill-based experimental sessions "there [was] so much entry that the average subject [lost] money in 34 out of 48 periods, and [earned] money in only four periods" (Camerer and Lovallo 1999, 315). So over-confidence and reference-group neglect really matter, and Camerer and Lovallo (1999) link those findings to the high outside-laboratory failure rates of new businesses.

Given those *high business failure rates* measured in outside-laboratory markets together with the laboratory-based explanation by Camerer and Lovallo (1999), what should be the *advice* to a young entrepreneur who just considers market entry with some innovation? What about the following advice: "Watch out, do not be 'overconfident,' do not overestimate your chances and abilities; statistically, too many entrepreneurs fail. Rather play it safe and go for a job in the management of a large company." But is this advice really helpful, is its theoretical basis even correct in the clustered-minds multiverse? Alright, there must be *some* low-consciousness versions (however, not *very*-low-consciousness versions) that do experience failure in some realities, otherwise neither the non-laboratory failure rates nor the results by Camerer and Lovallo (1999) would be possible. But, and that is the perhaps perplexing point I want to make here, there is enough 'room,' provided by the existence of *parallel markets*[14] and parallel minds clusters for many entrepreneur's *core consciousness* (versions of the entrepreneur where a lot of consciousness resides) to experience a successful market entry. Moreover, 'hammering' the fact that there is overconfidence out there, that there is reference group neglect and so on, might make it a self-fulfilling prophecy (see chap. 11) for young entrepreneurs to fail; so that an entrepreneur (perhaps unconsciously) opts for allocating a lot of consciousness into 'failure' realities and drawing away consciousness from success realities, clearly an undesirable way of conducting a vectorial choice (unless the respective individual wants to experience failure, for some of the reasons described in chaps. 9 and 11). So, in a way, overconfidence is a tricky phenomenon in the multiverse, and its interpretation as well as advice based on it might be chosen with care.

[14]It is clear that parallel realities must lead to parallel markets, side-by-side with many other things such as parallel New York Cities, parallel Berlins and parallel Notre Dame Churches.

Playing against Computers: A Different Story from the Perspective of the Multiverse?

Simultaneous market entry could be demonstrated to be a rich paradigm for linking economics and the multiverse. But I am sure that other games and markets offer a lot of potentially challenging as well as inspiring results, too. In the following, I will not endeavor the analysis of some other games in their own right or look at different types of markets in detail; this will all have to be left for future research. However, I will look at some other games here, albeit from a specific angle: I am going to compare the *play against computers* and the *play against other humans* with those games that have been chosen for this comparison by the respective authors. Why should the comparison of play against computers with play against humans be important from the perspective of the multiverse?

I would like to argue that there are good reasons to be very excited about this comparison, indeed. Playing against humans invokes *restrictions* to the composition of joint realities such as avoidance of very-low-consciousness situations as well as, perhaps, the *meaningfulness* of clustering that was mentioned above. None of those restrictions might play any role in games against computers if computers are interpreted as 'mindless hulks' without any sense of meaning[15] to start with. Therefore, experimental results that vary the counterparts of an individual: computer or human, should demonstrate different behaviors of that individual in those two scenarios. (Sure enough, the play against a computer is a funny blend of a game against an intelligent counterpart and a game against nature.) Although a multiverse perspective was never applied to the interpretation of the findings from the respective experiments, those experiments do find pronounced differences between the play against a computer and the play against a human, and I hence felt it might make sense to have a closer look at those differences. Note that the actual *reason* for the observed differences might be debatable at times; indeed, the respective authors' explanations (as well as their hypotheses) generally differ from the explanations I would like to suggest. Sure enough, the following speculative thoughts are far away from any scientific conclusiveness at this point. But I hope that they are precise enough to inspire research on this matter.

One of the first studies comparing play against humans with play against computers might have been the one by Abric and Kahan (1972) on the *prisoner's dilemma game*. In this game, often called social dilemma, two individuals might *independently* opt for either a cooperative or a non-cooperative choice. In prisoner's dilemma, the payoffs to *both* players are larger if both players choose to cooperate than if both players choose not to cooperate. The situation becomes tricky, however, because of the *asymmetric* cases. If *one* chooses to cooperate and the other does not, the non-cooperator earns a *higher* amount than in the situation where both

[15]This implies that I suppose that computers have no consciousness; this might be debated, but not within the scope of this book. Also, it could still be that humans attach meaning to their interaction with computers. I am not going to explore this point here either.

cooperate, and the sole cooperator ends up with a payoff that is even *lower* than the payoff in the case where both choose not to cooperate. Thus, the non-cooperator is able to take advantage of the cooperator. Not surprisingly then, the *Nash equilibrium* in this game is (non-cooperate/non-cooperate), but the socially optimal outcome is (cooperate/cooperate). This situation is often called social contribution dilemma and its basic structure can be found in many parts of life (e.g., climate control agreements between countries, regulation of overfishing and team production). The 'puzzle' with respect to this game—from the perspective of normative game theory—is that whenever it is played among humans, many people cooperate despite the Nash equilibrium indicating no cooperation. The question in games again computers naturally is whether those results hold, or to what extent.

The study by Abric and Kahan (1972) involved a few more features I am not able to deal with here, but the main result is that *subjects cooperate more with humans* (55% of the time) than with computers (35% of the time). Whereas the authors suggest an explanation via "perception of an actual relationship, of an interaction with the other person" (129). I would like to suggest an explanation via *behavioral restrictions in the multiverse that pertain to people, not machines.* Or in other words, whereas the number of realities that could be created via a play of this game is principally identical in games against humans or computers, human consciousness might not be willing to share too many realities where the other player is taking advantage of us. Avoiding very-low-consciousness situations (or, perhaps, clustering in meaningful ways; see above) might restrict the possibility of playing non-cooperative most of the time.

Kiesler et al. (1996) ask the question whether "communicating with a computer partner change the feelings or norms people follow in deciding what to do" (48). In order to address this question, they also investigate a prisoner's dilemma game. Based on social-identity theory, their main hypothesis is as follows: "People will behave more cooperatively towards a human partner than towards a computer partner, but their cooperativeness with a computer will be more like that with a person, the more human-like features the computer partner has" (52). However, they earlier stated this: "A competing conception of people's interaction with technology is that this interaction is only social by mistake. One can imagine novice computer users erroneously treating a talking computer as though it were human because they lack knowledge of technology. Perhaps it is comforting to treat a computer as though it understands us" (49). Although they dismissed this line of arguments, I would like to keep it in mind as important, also given the time of the study in the first half of the 90s of the last century where people were in general less familiar with computers than they are now.

I am again abstracting from the various interesting features of the experimental design by Kiesler et al. (1996) and look at the basic findings, only. Even though there was some 'communication' taking place between the partners (even if the partner was a computer), behavior on first trials dramatically differed between computer-partner and human-partner conditions: "Eighty percent of the participants in the person condition cooperated, approximately twice the 41% who cooperated in the computer conditions" (56). Similar results occurred with respect to cooperation promises (in communication) being kept or not. People kept their promises more when playing against people than computers.[16] As already argued within the interpretation of findings by Abric and Kahan (1972), more behavioral restrictions seem to be applied in games against humans than machines.

In a study by McCabe et al. (2001) different, related games concerned with trust and reciprocity (trust game, punish game, mutual advantage game)[17] were played by (part of the) subjects in the brain scanner (fMRI). Interestingly, certain brain regions become only active with individuals that *do cooperate* and *only* if they are playing *against humans*: "The six subjects (…) with the highest cooperation scores show significant increases in activation in medial prefrontal regions during human–human interactions when compared with human–computer interactions" (11,833). Perhaps, this could be the brain region that is activated when behavioral restrictions in play with humans are evaluated in the multiverse. Looking from this multiverse-inspired perspective, however, it is unclear why individuals that chose *not* to cooperate would *not* have to check for restrictions by using the very same brain regions: "The six subjects who received the lowest cooperation scores (…) did not show significant activation differences in medial prefrontal cortex between the human and computer conditions" (11,833-11,834). This point has to be left open, here.

In a study by Rilling et al. (2004), respondents' brain reactions were again investigated using fMRI whilst playing two different games over multiple rounds, a prisoner's dilemma game and an *ultimatum game*. I have already explained the prisoner's dilemma game. The ultimatum game has been introduced to the literature by Güth et al. (1982). In this two-player game, the proposer gets an amount of money by the experimenter, say, $10.00, and is supposed to propose any part of it as an offer to the responder. The responder may either accept or reject that offer. If the offer is accepted, the responder gets the proposed amount of money, and the responder keeps the rest. If the proposal is rejected, the entire amount of money is lost; the responder does not get any money, and the proposer loses his endowment. The normative, game-theoretic prediction is that the proposer proposes a cent and the responder accepts that proposal. Empirical data, however, demonstrate that low proposals are rejected (for reasons of fairness or inequity aversion), and average offers are far beyond the theoretical cent (often ranging up to 50% of the initial endowment). In Rilling et al.'s (2004) experiments, "(…) rejection rates for 7:3,

[16]There was some strange pattern occurring in first trials as well as kept promises. People disliked the human-like computers (face, voice) and clearly preferred the treatment were the computer just sent text messages. I would like to argue that this might be due to the text communication actually being more natural somehow than an artificial face or voice. Only looking at the text communication treatment, the difference between computer and person as counterparts became quite small but did not completely disappear.

[17]Unfortunately, the authors were precise only with respect to the trust game. In a trust game, one player goes ahead with transferring an amount of money to a second player, and that player may reciprocate by returning part of the transferred amount that will in turn be multiplied with a certain factor. The game seemed to have been played in a simplified version. Details on the other two games are totally missing. It might be speculated that the mutual interest game has been some form of a prisoner's dilemma game, but one cannot be sure.

8:2, and 9:1 offers from human partners were 5%, 47%, and 61%, respectively. Rejection rates for 7:3, 8:2, and 9:1 offers from computer partners were 5%, 16%, and 34%, respectively" (1698). Thus, humans were *expected* to use a *smaller range* of the possible offers than computers (consistent with the 'restrictions in multiverse' hypothesis) and rejected more frequently if they violated that expectation. Also, in the prisoner's dilemma game participants cooperated with human partners in 81% of the cases but only in 66% of the cases with computers. With respect to the fMRI findings, the ultimatum game played against computers hardly activated any brain region that the authors considered relevant for the choices to be made, whereas the ultimatum game played against humans did. In the prisoner's dilemma game, the brain reactions to humans versus computers were less different but still showed significantly more activation in certain regions in play against humans rather than computers consistent with the above hypothesis that some brain regions might be used to analyze behavioral restrictions in the multiverse with human counterparts.

Concluding Remarks

The current chapter tried to link strategic games and economic decisions with the framework of the clustered-minds multiverse. The first question that might come into mind is why I did not make any use of a very recent monograph that has been mentioned already in chap. 1 of this book and that carries the title *Quantum Economics*. However, Goswami's (2015) book is solely concerned with the important re-development of *idealistic thinking* (as noted in chap. 1) and is rather related to the monograph by Eisenstein (2011) on *Sacred Economics*. It does not contain any specific theoretical developments regarding a closer connection between quantum mechanics and economics—not to mention the multiverse interpretation—such as those suggested in this chapter.

I have to admit that the development in this chapter had to remain inconclusive for several reasons. E.g., I have mainly dealt with one specific game or economic decision situation: simultaneous market entry, and I have left out reallocations of consciousness and the restriction of moderate allocations of consciousness. Regarding the latter two aspects, given the already complex analysis presented, this would have made the deliberations in this chapter intractable. But clearly, this would have added relevant behavioral aspects regarding chances and limitations with strategic decisions and should thus be looked at in future research. Regarding the prior, market entry is a paradigm that I judge as being quite fruitful for the discussion, and I hope I was able to prove this opinion to be reasonable in the course of this chapter. I have explored this paradigm in the versions of Rapoport (1995) and Camerer and Lovallo (1999). Each of those paradigms allowed for important connections with multiverse thinking: e.g., meaningfulness of clustering as a different 'story' for Rapoport's (1995) findings as well as overconfidence/over-entry as potential artifacts of a singular reality analysis. The question is how some of those thoughts could be substantiated, empirically, and be generalized to other markets and strategic situations, to other behavioral effects etc. A better understanding of meaningful clustering beyond a different 'story' is especially important for future research, not only within the context of this chapter but for developing quantum decision making more generally.

The comparison of strategic play against computers with strategic play against humans revealed a number of interesting behavioral differences including different brain reactions. It is quite clear that those results might generalize to other decision situations. First, several different games were analyzed (with some focus, however, on prisoner's dilemma), second, the results were general enough—I would like to argue—to hold for several setups in game theory and economics. The open question is whether the tentative as well as speculative explanation I have offered for the behavioral differences (computers as counterparts are located somewhere between games against nature and games against humans; and games against computers lead to fewer restrictions with respect to 'feasible realities' than games against humans) is better (more appropriate) than the several explanations offered by the respective authors or at least contributes to the explanation of those differences in play. It is certainly hard—but perhaps not impossible—to come up with research designs that test those explanations against each other. This is another nice challenge for future research.

PART IV
CONSEQUENCES FOR SELECTED PSYCHOLOGICAL PHENOMENA AND EXPERIMENTAL RESEARCH IN THE SOCIAL SCIENCES

Chapter 11
Repetition Compulsion and Self-Fulfilling Prophecies

Repetition Compulsion—"Beyond the Pleasure Principle"

A famous small monograph by Freud (1955 [1920]) is carrying as its title the last part of the title of this section. In this publication, Freud struggles with the 'pleasure principle' of psychoanalysis. He identifies the pleasure principle with an "'economic' point of view" (Freud 1955 [1920], 7); and I would like to underline here that this principle is indeed closely related to the concepts of utility dealt with in Chaps. 8 and 9. Not surprisingly, then, Freud faces the same type of problems I was facing in Chap. 9 when trying to understand why people might *choose* to make all kinds of odd and unpleasant experiences (for instance in movies they watch in the cinema or at home).

The example Freud (1955 [1920]) is looking at is a compelling one when one tries to find counterexamples for individuals following any 'pleasure principle' or utility maximization: *repetition compulsion*, a phenomenon he perceives as relevant not only with neurotic individuals but also with regular people:

> What psychoanalysis reveals (...) [with] neurotics can also be observed in the lives of some normal people. The impression that they give is of being pursued by a malignant fate or obsessed by some 'daemonic' power; but psychoanalysis has always taken the view that their fate is to the most part arranged by themselves and determined by early infantile influences. The compulsion which is here in evidence differs in no way from the compulsion to repeat which we have found with neurotics, even though the people we are now considering have never shown any signs of dealing with a neurotic conflict by producing symptoms. Thus we have come across people all of whose human relationships have the same outcome: such as the benefactor who is abandoned in anger after a time by each of his *protégés*, however much they may otherwise differ from one another, and who thus seems doomed to taste all the bitterness of ingratitude; or the man whose friendships all end in betrayal by his friend; or the man who time after time in the course of his life raises someone else into a position of great private or public authority and then, after a certain interval, himself upsets that authority and replaces him by a new one; or, again, the lover each of whose love affairs with a woman passes through the same phases and reaches the same conclusion. (Freud 1955 [1920], 22)

© Springer Nature Switzerland AG 2018

C. D. Schade, *Free Will and Consciousness in the Multiverse*,
https://doi.org/10.1007/978-3-030-03583-9_11

In his publication, Freud (1955 [1920]) then even invokes a *death instinct* and some "urge (…) to restore an earlier state of things" (30) as explanations. When Freud began publishing about repetition compulsion, however, he had seen repetition compulsion as a means to finally achieve *mastery* over the conflicts and traumata involved, and many have rather resonated with this first idea (see Levy 2000 and the literature referenced there) although mastery is rarely achieved (Chu 1991; van der Kolk and Greenberg 1987). Unfortunately, this interesting 'dispute' is not of the kind I might be able to resolve, here.

Since I have already dealt with the problem of defining and interpreting utility in Chap. 9 and I am not able here to provide a new perspective on this matter, I am not going to go into this problem in more detail. The existence of repetition compulsion simply adds more evidence to what has already been discussed in Chap. 9 under the heading of individuals 'choosing odd and painful experiences,' and it shows again how urgent and difficult is the correct definition of (experienced) utility and understanding the potential interdependencies, in terms of total utility, of the set of realities chosen by an individual. Repetition compulsion is finally relevant to, again, shed some light on the equally puzzling question as to why individuals do not allocate consciousness partially *away* from realities that turn out to be painful (even if there are restrictions, presumably relevant for both vectorial decisions as well as reallocations of consciousness, such as the avoidance of very-low-consciousness or extreme-allocations-of-consciousness situations, the most painful repetitions should be avoided, if positive utility is looked for). Perhaps the same reasons are relevant for such non-reallocation decisions than for choosing in favor of painful realities to start with.

What I want to draw the attention of the reader to in this chapter from now on is *not* the question as to *why* people are engaging in repetitions, i.e., what their exact motivation is, but rather to the—from my perspective—even more puzzling question as to *how they manage to co-produce the experience of those repetitions*. It is again Freud (1955 [1920]) who first wondered about this implicitly by discriminating between two cases:

> This 'perpetual recurrence of the same thing' causes us no astonishment when it relates to *active* behavior on part of the person concerned (…). We are much more impressed by cases where the subject appears to have a *passive* experience, over which he has no influence, but in which he meets with a repetition of the same fatality. There is the case, for instance, of the woman who married three successive husbands each of whom fell ill soon afterwards and had to be nursed by her on their death-beds. (Freud 1955 [1920], 22)

I have to admit that I am already impressed by the first case described by Freud (active behavior), because it shows that individuals might, at least partially and clearly unconsciously, be able to select their environment to an impressive extent. But I agree that the second case (passive experience) is even more impressive. Unfortunately, Freud does not say much on the—from my perspective quite necessary—explanation as to why such passive influences are possible at all.[1]

[1]This statement still holds in modern psychoanalysis.

The clarification of this phenomenon within the framework of the clustered-minds multiverse will hence be in the center of the next section. In the subsequent section, I will turn to a somewhat related phenomenon, self-fulfilling prophecies, define it, provide a plethora of examples and craft an explanation of this phenomenon within the multiverse framework as well. Another section will concern itself with the question whether multiple equilibria (that have already been dealt with in a different context in Chap. 10), offered as an explanation for the workings of self-fulfilling prophecies within economics, can be seen as an alternative explanation to the multiverse account of this phenomenon. Concluding remarks will finish the chapter.

Repetition Compulsion as Vectorial Choice

The happening of at least some of the cases of repetition compulsion is hard to comprehend in a singular universe: E.g., how is it possible that the woman in Freud's example marries three husbands in a row that would all, soon after the marriage, die? And how is it possible that a man makes friends with different individuals, one after another, with all friends being willing and able to betray him? I would like to argue that there are simply not enough *degrees of freedom* within a singular reality (the entire course of events would also have to be seen as predetermined in a deterministic singular reality) for those things to be happening. Why shouldn't there be at least one healthy husband among all the deadly ill and why not at least one man that finds it unethical to betray his friends? The multiple repetitions are intuitively *implausible* in a singular reality. Does the clustered-minds multiverse help with their explanation?

In Chap. 8, different drivers of vectorial choice have been considered. They are all relevant to understanding the phenomenon of repetition compulsion. Note that this means that I will consider a Freudian repetition a 'regular' vectorial choice, but certainly not to trivialize the tragedy arising for the respective individuals by some of the painful repetitions. However, what I then have to explain is *how* large parts of consciousness are allocated to realities where repetitions occur. How is that allocation exactly happening, in the context of repetition compulsion? Let me go over the parameters stated in Table 8.4, again, and apply all those factors to the phenomenon at hand.

The first driver that was mentioned in Table 8.4 is *unconscious motives of the decision maker*. And this clearly is already the most important driver of repetitions. Table 8.4 was also explicit about the genesis: "preferences within the person, but unknown to him, hence not explicit part of decision models." The man who has repeatedly been betrayed by friends might clearly not be aware of the fact that he is, unfortunately, unconsciously working in favor of the next disaster. The impact of the factor 'unconscious motives …' was considered to have a potentially substantial impact if the unconscious drivers are stronger than those that are consciously analyzed, undermining seemingly rational judgements. One could imagine that this applies in the context of the experiences of the gentleman in Freud's analysis.

The last aspect mentioned with respect to the unconscious motives in Table 8.4 sounds as if it had already been developed within psychoanalysis (e.g., Anna Freud 1937; Friedman 2019): "The more the unconscious motives are known to the decision maker, the more they can explicitly be integrated into decision models (or those models 'debiased')." Given the fact that the 'treatment technique' of psychoanalysis aims at making unconscious conflicts (partially) accessible to the conscious, repetition compulsion describes a situation where the most important drivers of an individual's behavior have not been made conscious so far. The multiverse-specific novelty is that the unconscious is able to co-decide as to how much emphasis to put on each reality out of a set of many realities, some allowing for the experience of a repetition, whereas in a singular reality, such a vectorial choice is impossible.

The impact of *other, including 'non-human' factors* (the normal 'course of the world,' including macro probabilities to handle a decision maker's ignorance) might exclude a repetition from occurring. If simply no one is available as a friend at some point, the basis for allocating a lot of consciousness into a 'betrayal reality' is not provided for the gentleman in Freud's analysis. *Unknown motives of other decision makers (conscious and unconscious)*, "restricting the allocation of consciousness to certain realities when operating close to the fringe of a consciousness cluster," might become relevant in the context of repetition compulsion if a new friend is found but only very-low-consciousness versions of him are willing to engage in betrayal. In this case, allocating a lot of consciousness into a repetition is, again, not possible at this point, albeit for a different reason. Note that I do not think that those two factors are *often* restrictive with the case at hand; or in other words, the multiverse-specific 'advantage' of being able to allocate a major part of consciousness to repetition realities, described within the vectorial implementation of unconscious motives above, is only loosely restricted by the two factors just described.

Quite complex and also very speculative with respect to their effect are the potential *utility interdependencies between different versions of the individual* as well as potential *variety-seeking tendencies across versions of the individual*, maximizing differences between their experiences (versus rather concentrating consciousness on a few, related realities, restricted, however, by the constraint of 'looking for moderate allocations;' see Chap. 4, list of premises of the clustered-minds multiverse). In any case, if the gentleman in Freud's example has allocated some consciousness into realities with reliable friends, already, utility interdependencies and variety-seeking tendencies might either hinder or stimulate allocating a lot of consciousness into betrayal realities. The same applies to decisional conflicts that are, anyway, related to utility interdependencies.

Free will appears as a *residual* with only few power if all the discussed factors point in the same direction, 'organizing' the next betrayal, and with the individual getting hardly any conscious knowledge of it. (To be sure, using the word 'organizing,' by no means I want to be cynical or suggest that individuals are fully responsible for all those, often painful repetitions happening in their subjective experience. It is, anyway, unclear as to how much individuals should be held responsible for their unconsciously driven actions as has been discussed in Chap. 7;

see also Herdova 2016). However, free will might gain some ground to, e.g., work against the next betrayal, if either the discussed factors are partially in conflict or if many unconscious factors are already known to the individual.

Self-Fulfilling Prophecies: Definition, Examples, and Explanation via the Clustered-Minds Multiverse

Self-fulfilling prophecies are an important concept occupying the remainder of this as well as the next chapter. They are a *frequent phenomenon* as will be demonstrated via numerous examples, but they are—as was the case with repetition compulsion—hard to explain within a singular reality. There are two exceptions so far to the general shortage of theoretical explanations. Self-fulfilling prophecies might be seen as consistent with *constructivism* (to be looked at towards the end of this section) and with the *existence of multiple equilibria* (an economic concept that has been introduced in the last chapter and that will be looked at as a potential explanation in the next section). Both explanations might be seen as lower-level descriptions (or special cases) of the more general concept of the clustered-minds multiverse providing the *physical basis* for those explanations. In fact, the minimum that might be said is that the clustered-minds multiverse *completes* the explanation of self-fulfilling prophecies.

What is a self-fulfilling prophecy?[2] "(…) public definitions of a situation (prophecies or predictions) become an integral part of the situation and thus affect subsequent developments. This is peculiar to human affairs. It is not found in the world of nature" (Merton 1948, 195). According to Merton (1948, 193), the idea of self-fulfilling prophecies might be traced back to the *Thomas theorem*: "If men define situations as real, they are real in their consequences" (Thomas and Thomas 1928, 571–572). The definition by Watzlawick (1984a) is more general than that by Merton: "Self-fulfilling prophecy is an assumption or prediction that, purely as a result of having been made, causes the expected or predicted event to occur and thus confirms its own 'accuracy'" (Watzlawick 1984a, 382).

Many examples show the workings of self-fulfilling prophecies in *social interaction contexts*. E.g., Merton (1948) analyzes the antecedents and consequences of a hypothetical *bank run*. A bank's reserves are shrinking quickly if most clients want to withdraw their money on a certain day. Clients will do so if they expect economic problems of the bank. Hence, *it is the expectation of the bank's bankruptcy that will cause the bank to be bankrupt*. Merton's (1948) main example are then the dynamics of *racial discrimination*. Rist (1970) analyzes how *social*

[2]The usual definition of self-fulfilling prophecies, to be found on the same page, is the following: "The self-fulfilling prophecy is, in the beginning, a false definition of the situation evoking a new behavior which makes the originally false conception come true" (Merton 1948, 195). I opted for a different one because with parallel realities, there might not be a principal difference between right and wrong conceptions of a singular one.

class influences expectations by teachers, how those teachers' expectations influence how children are treated, and how their performance will finally stay closely related to social class membership. And a field experiment by Rosenthal and Jacobson (1968) demonstrates that a randomized manipulation of teachers' expectations towards certain students results in substantial performance differences between those and other students.[3] Since there are many more examples on "the power of social beliefs to create reality" (Jussim and Eccles 1995, 99; see also the literature referenced there as well as the last footnote) one might wonder as to why this is possible: "(…) are people so *malleable* that they readily fulfil other's erroneous expectations?" (Jussim and Eccles 1995, 74; italics mine). One might feel reminded of the quite similar question that was asked with respect to repetition compulsion above.

But before I go into the explanation of the phenomenon, it is instructive to look more deeply into some of the exciting outside-laboratory situations reported and partially discussed in Rosenthal (1976) and Watzlawick (1984a). Rosenthal (1976, 129) mentions a striking example of self-fulfilling prophecies in a study by Whyte (1943) where a group of young men would regularly meet for bowling and would 'know' beforehand who would bowl well on a certain evening. Typically, especially the leaders' predictions were quite accurate. Rosenthal (1976, 129) speculates about a possible explanation via reduced anxiety on the side of those young men whose performance was predicted to be good on a certain evening, and higher anxiety on the side of those young men whose performance was predicted to be bad. Another anecdote is to be found in Guthrie (1938) where a "shy, socially inept young lady became self-confident and relaxed in social contacts by having been systematically treated as a social favorite," (Rosenthal 1976, 130) because a group of college students had manipulated the behavior of her environment. A large-scale societal (or economic) development, quite similar in structure to the thought experiment on bank runs by Merton (1948, see above), but that actually took place, was mentioned by Watzlawick (1984a): "In March 1979, when the newspapers in California began to publish sensational pronouncements of an impending, severe gasoline shortage, California motorists stormed the gas stations to fill up their tanks and to keep them as full as possible. This filling up of 12 million gasoline tanks (which up to this time had on the average been 75% empty) depleted the enormous reserves and so brought about the predicted shortage practically overnight. (…) After the excitement died

[3]In a well-known Meta study by Rosenthal and Rubin (1978) on *interpersonal expectancy effects*, the above examples would fall under the category of *everyday situations* and are discriminated from all kinds of laboratory situations concerned with *reaction time* or *animal learning* to be dealt with in the next chapter. The studies on everyday situations exhibit a large variability of outcomes, but also reveal large effects sizes, on average, i.e., the self-fulfilling prophecy effects are strong. Only laboratory studies on animal learning and psychophysical judgments exhibit even stronger effects (see for those comparisons Rosenthal and Rubin 1978, 380, especially Table 3). The mean effect size of the studies on everyday situations is 0.88 in Rosenthal and Rubin (1978). The classification of this effect size as being strong follows the classification by Cohen (1977).

down, it turned out that the allotment of gasoline to the state of California had hardly been reduced at all" (383).

Even more important for the point to be made in this chapter are some of the examples around deaths and illnesses that Watzlawick (1984a) covers. One example is psychiatric illnesses that, according to Watzlawick (1984a, 388) and others (ibid.; see the literature referenced there) might have a self-fulfilling effect: "(…) an essential part of the self-fulfilling effect of psychiatric diagnoses is based on our unshakable conviction that everything that has a name must *therefore* actually exist" (Watzlawick 1984a, 388; italics mine). Another example is "Voodoo Death" (Watzlawick 1984a, 388): "'Magic' diagnoses, in the actual sense of the word, have of course been known for a very long time." Building up upon Cannon (1942), Watzlawick (1984a) comes up with a number of examples that *he* interprets in the way that an aboriginal individual in certain parts of the world who knows that he has been cursed by a powerful person such as a medicine man is "helpless against his own emotional response to this death sentence and dies within hours" (388). Based on Simonton and Simonton (1975, 31), Watzlawick also sees components of self-fulfilling prophecies in the etiopathology of cancer, specifically, in "(…) the belief system of the patient, that of the patient's family, and, third, that of the attending physician. That each of these belief systems can become a self-fulfilling prophecy seems credible" (Watzlawick 1984a, 389). An interesting question that Watzlawick (1984a) also draws his attention to in this regard is how "much can and should a physician tell his patients, not only about the gravity of their illnesses, but also about the dangers inherent to the treatment itself? (…) Since in the patient's eye a doctor is a kind of mediator between life and death, his utterances can easily become self-fulfilling prophecies" (389–390). He then turns to the "potentially healing effect of positive predictions," autosuggestions, and hypnotherapeutic interventions, to "influencing the course of an illness (…) by positive imagery," all these being cases of self-fulfilling prophecies (Watzlawick 1984a, 390). He finally sees the *effectiveness of placebos* as actual cures of diseases as examples of self-fulfilling prophecies (Watzlawick 1984a, 390–391): "(…) the claim of the doctor who administers the placebo that it is an effective, newly developed medicine and the patient's willingness to believe in its effectiveness create a reality in which the assumption actually becomes a fact" (ibid.).

Let me start the interpretation of self-fulfilling prophecies in the multiverse with three especially insightful references, again by Watzlawick (1984a):

> Everyday experience tells us that only few prophecies are self-fulfilling (…): Only when a prophecy is believed, that is, only when it is seen as a fact that has, so to speak, already happened in the future, can it have a tangible effect on the present and thereby fulfill itself. (385)

> The fact that we are responsible to the world in its entirety and to a much higher degree than is dreamed of in our philosophy is for the present almost unthinkable. (387)

> Self-fulfilling prophecies are phenomena that not only shake up our personal conception of reality, but which can also throw doubt on the world of science. They all share the obviously reality-creating power of a firm belief in the "suchness" of things. (391)

Sure enough, Watzlawick (1976, 1978, 1984b) would argue that the theory for all this is *constructivism*; but there is a close association, I would like to argue, between constructivism (in its various versions) and the clustered-minds multiverse proposed in this book, based on quantum mechanics[4]: Both approaches would argue that subjective reality is influenced by the observer (see, e.g., for the version of radical constructivism, Glasersfeld 1984). Specifically, if consciousness is able to put primary emphasis on a subset of realities (and direct awareness away from other realities), realities that are *expected* can be made *more important* within consciousness of the observer; or more precisely, that version of the individual residing in a reality that is consistent with expectations gets a much larger share of consciousness than those versions residing in realities that are inconsistent with the 'prophecy.'

This all seems to be related to the explanation that was offered for repetition compulsion, and therefore it is unnecessary to again walk through all the aspects described in Table 8.4 as drivers of vectorial choices. However, a potential difference between the explanation of repetition compulsion and the explanation of self-fulfilling prophecies is the role of conscious versus unconscious expectations, whereas the prior are closely associated with free will. It is clear that repetition compulsion is predominantly unconscious (see above), whereas the case for either conscious or unconscious 'action' is harder to be made with self-fulfilling prophecies. Some of Watzlawick's above references might easily be misunderstood as self-fulfilling prophecies having to be *conscious*; this at least being the case if (firm) *beliefs* are seen to require conscious thought. I disagree, however. I would rather argue that firm beliefs might be conscious beliefs *or* unconscious beliefs *or* a mixture of both. Indeed, many of the above examples might be associated with such mixtures.

Multiple Equilibria as an Alternative Explanation for Self-Fulfilling Prophecies?

There is a stream of literature in economics, closely related to the case of multiple equilibria in game theory (sometimes explicitly based on game theory), that might be seen by game theorists and economists as offering an alternative explanation for the effectiveness of self-fulfilling prophecies. However, it is clear that economic explanations only pertain to a *subset* of self-fulfilling prophecies. The cure with a

[4]Although constructivism is not explicitly based on either quantum mechanics or even the multiverse interpretation of it, Watzlawick was aware of the similar spirit of a subjective interpretation of quantum mechanics and constructivism; this becomes apparent in the epilogue to his edited volume "The invented reality" (Watzlawick 1984b, 330–331) where he references and endorses Schrödinger's perspective provided in "mind and matter" (2004 [1958]).

placebo, for instance, is not among them.[5] Moreover, I will be analyzing the exact status of an explanation via multiple equilibria in light of the clustered-minds multiverse below (see also Chap. 10).

Whereas a complete overview of the relevant economic papers is impossible within the scope of this book, I would like to nevertheless give an idea of two different applications. One is macroeconomic theory, concerned, e.g., with the question as to what predicts economic growth. It can formally be shown that growth is partially dependent on *expectations* in the sense that they are self-fulfilling (Farmer and Woodford 1997); the formal requirement is the existence of *multiple equilibria*, i.e., of multiple possible market outcomes where actual growth and expectations are consistent with each other.[6]

The other case that I would like to spend a little more time with is the relationship between technology adoption and critical mass (the analysis of this relationship would be classified by an economist as belonging to the field of industrial economics). A working paper by Li and Zhou (2015) concerns itself with the important question whether the U.S. electric vehicle market is a multiple equilibria situation.[7] The vehicle market is characterized by a certain feature that economists would call *positive indirect network externalities*. "In markets with positive indirect network effects, one side of the market tends to wait for the other side to act before taking its own action. Previous literature (…) has emphasized the multiple equilibria issue and the "chicken-and-egg" coordination problem" (Li and Zhou 2015, 2; see also Caillaud and Jullien 2003).

The adoption of an electric vehicle by a consumer critically depends on his expectation of the spatial density of public charging stations in the future; whereas investor's willingness to build public charging stations critically depends on the expected number of electric vehicles bought by the U.S. consumers (clearly, the same holds for other countries in the world). It is easy to see that many equilibrium points can be reached, based on different expectations. One equilibrium could be, e.g., characterized by an expectation of low investments in public charging stations. If this can be predicted or anticipated, low purchases of electric vehicles arise. This situation leads investors expecting the building of public charging stations as not profitable. Building small numbers of electric vehicles will not be profitable either, etc. An equilibrium point below critical mass is reached. The technology 'flops.'

[5]The determination of exact boundaries between cases that might and might not be explained via multiple equilibria is beyond the scope of the book.

[6]For a sophisticated mathematical analysis of the general problem of self-fulfilling equilibria see Mas-Colell and Monteiro (1996).

[7]Li and Zhou (2015) are, however, more interested in the determinants of equilibria such as the nature of indirect network effects, consumer preferences and the like. I am taking the leeway here to use their example of electric vehicle adoption in the U.S. but write the story more around a 'pure expectations' setup. This can be done in a multiple equilibrium situation but is not exactly what these authors have had in mind.

Now let me consider an alternative equilibrium. Here, consumers expect a high number of public charging stations. Their utility of buying an electric vehicle is larger now than in the situation with only few public charging stations, hence the rate of adoption of electric vehicles is higher. If this is anticipated by investors, more public charging stations will in fact be built etc. Thus, consumers will be happy with their vehicles, producers of the vehicles will have a high profitability because of the large numbers produced, and the investors will be happy because their public charging stations will be used. An equilibrium point above critical mass is reached. The technology 'booms.'

Unlike Li and Zhou (2015) that look for *economic determinants* of different equilibrium outcomes and analyze measures the *government* can take to move behavior to an above-critical-mass equilibrium such as purchase subsidies for electric vehicles or income tax cuts, I have presented this analysis for a different reason. In the clustered-minds multiverse, different equilibria might coexist: There might be realities where the critical mass is surpassed and others where it is not. The question is where *consciousness* of different individuals will *reside* to a higher and to a lower extent. Note that I have talked of expectations being decisive for different equilibria to arise. Expectations are a phenomenon of consciousness, so indirectly, the fact that realities might become more or less relevant (because a lot of consciousness resides there) is somehow consistent with the economic idea of expectations.

So, the remaining question here must be what the relationship is between equilibrium analysis and multiverse theory. Is equilibrium analysis an alternative theory to the multiverse account of the phenomenon? The last chapter has explored the relationship between equilibrium analysis and the clustered-minds multiverse for simultaneous market entry. As in the last chapter, however, the number of feasible equilibria (as well as of off-equilibrium outcomes that are in principle possible) might be limited by the very-low-consciousness restriction as well as by what was called meaningful clustering. Concentrating here, for the sake of simplicity, on the prior, individuals stay away from situations where they are almost *alone*, surrounded by very-low-consciousness entities. The last chapter has demonstrated how complex the considerations already are in simultaneous market entry with a relatively small number of players. In a large numbers situation such as electric vehicle adoption with heterogeneous players involved (state, consumers, investors etc.), it is unclear, given the knowledge we currently possess, what exactly are the limits to the number of *possible, strong, high-consciousness* universes; or at least how many equilibria (and non-equilibrium outcomes) do actually coexist with some, *above minimal*, amount of consciousness allocated to them.

It is also unclear how strong the impact of consciousness reallocations can be in such a situation with, say, millions of individuals being involved. Is it possible to partially remove consciousness from an equilibrium with many involved individuals; is this easier or more difficult than removing it from an equilibrium with only few individuals involved? On the other hand, is it actually simple to move consciousness ex post into an equilibrium where a lot of consciousness resides? And is there a substantial difference between those ex post allocations and economic decisions (i.e., new vectorial choices within an economic context) that simply bring a certain version of the individual into that equilibrium such as buying an electric vehicle when 80% of the population already bought one? These are interesting questions to be addressed in future research.

But one thing *is* very clear. Equilibrium analysis cannot be the *full* story; hence it cannot replace considerations based on the clustered-minds multiverse. Equilibrium analysis might support the multiverse account in better understanding some of the drivers relevant for vectorial choices (in the sense of 'other factors' according to Table 8.4), but it is not an alternative explanation. The economist predicts that despite the fact that there are multiple equilibria, individuals will finally pick only *one equilibrium outcome* (and make their strategy choices accordingly) via so-called equilibrium refinements (see, e.g., Harsanyi and Selten 2003; this type of thinking being somewhat related to collapse interpretations of quantum mechanics, see Chap. 2, but with a very different theoretical reasoning underlying). The clustered-minds multiverse will always lead to a plethora of equilibrium (and non-equilibrium) outcomes coexisting, even though the exact number of possible worlds is highly situation-dependent, especially hard to predict in situations that are as complex as the one just analyzed.

Concluding Remarks

This chapter translated repetition compulsion as well as outside-laboratory cases of self-fulfilling prophecies into the framework of the clustered-minds multiverse. It should be noted that on the one hand the *concept* of repetition compulsion has been critically discussed by some quite early (e.g., Kubie 1939), but on the other hand, there is continuing interest in the concept (e.g., Levy 2000; Kitron 2003). The interest might be due to the fact that the *phenomenon* of repetition compulsion is ubiquitous. The interest in self-fulfilling prophecies seems to have peaked in the seventies and eighties of the last century and then slowly declined—with the notable exception of economics as has been shown in the last section. However, there have also been a couple of novel studies on self-fulfilling prophecies outside economics, especially with respect to experimenter effects—the other important part of self-fulfilling prophecies—to be dealt with in the next chapter. Perhaps, the scientific interest in both repetition compulsion and self-fulfilling prophecies—outside economics—has been suffering from a slow progress in theory. The problem with both concepts has been their *explanation*.[8]

Repetition compulsion is a concept from psychoanalysis, actually an important one because the phenomena of *transference* and *countertransference* that result

[8]I do not mean to downplay the important achievements made in psychoanalytic theory and in the theory of constructivism. But I would like to argue that the clustered-minds multiverse is a theory with larger generality, in a way comprising the psychoanalytic and the constructivist' theories.

from it are main ingredients of the treatment technique of psychoanalysis.[9] But psychoanalysis finds it, firstly, hard to offer a sufficient explanation for individuals' search for 'negative utility' and, secondly, should be concerned about the fact that a large *plasticity* of the environment has to be assumed that is implausible in a singular universe. A theoretically sound explanation of self-fulfilling prophecies also seems to have been hard so far, with the exception of those self-fulfilling prophecies that can be described as multiple-equilibria situations, mostly in economics. Constructivism is not a theory that *explains* the phenomenon of self-fulfilling prophecies in narrow terms, it is rather a framework that describes its consequences (and it describes them quite similarly to the description offered within the clustered-minds multiverse). Thus, a lack of convincing theory might have been a valid description not only of the situation with many self-fulfilling prophecies outside the laboratory, but also for the large number of cases (and the huge effect sizes) to be found in the laboratory (see Chap. 12). But I suppose that this situation can be altered.

The reason is that I feel that this chapter has achieved more than just showing consistency of the clustered-minds-multiverse perspective and the two analyzed phenomena: repetition compulsion and self-fulfilling prophecies. Actually, the multiverse perspective is able to draft an explanation for them—to be enhanced in future research—and thus adds *scientific credibility* to those exciting and important parts of our reality in turn. Specifically, the multiverse perspective is able to explain the second part of the supposed agenda of theoretical development in psychoanalysis with respect to the repetition compulsion phenomenon, the plasticity of the environment. It is also able to give a convincing account of the mechanism of self-fulfilling prophecies. And finally, even though economists do possess a formal instrument to treat a subset of self-fulfilling prophecies in their domain, multiple equilibria analysis, I was able to show that this instrument is (at best) incomplete; that the clustered-minds multiverse is in principle needed to understand restrictions to the multiplicity of equilibria (and off-equilibrium behavior) on the one hand, but also to understand the parallel existence (not in a hypothetical, but in a real sense of the term 'existence') of many of them on the other hand.

---

[9]According to the theory of transference (e.g., Racker 2001), the patient lives through early-childhood traumata, say, with his mother, by unconsciously perceiving the psychotherapist as behaving *as the mother did*. And the trained psychotherapist should reflect this back in certain ways and heal the patient's trauma in turn. Countertransference (e.g., Racker 2001; for a historical overview: Stefana 2017) is a phenomenon with two sides: unconscious emotional reactions (i.e., repetitions) of the therapist directed towards the client that are often seen as a problem of the therapist, but also diagnostic countertransference, where the therapist is able to interpret his own feelings in favor of the patient's treatment. The challenge for the therapist is to disentangle his feelings towards the client into those two components.

Chapter 12
A Generalized Rosenthal Effect in Experimental Research in the Social Sciences

A Macro-World Measurement Problem?

The last chapter dealt with repetition compulsion as well as with those self-fulfilling prophecies that play a role in social interaction *outside the laboratory*. An important (as well as special) case of social interaction that was excluded in that chapter is *laboratory experimentation* in the social sciences, e.g., psychology, sociology and economics. The self-fulfilling prophecies relevant for this case of social interaction: *experimenter expectancy effects*, will be at the core of the current chapter. This type of interpersonal expectancy effects is also a special case of the more general class of experimenter effects including personality characteristics of the experimenter etc. (Rosenthal 1976; Rosenthal and Rubin 1978).

In fact, much of the early research on self-fulfilling prophecies has been concerned with experimenter expectancy effects in the social sciences, i.e., with the effect that researchers' expectations might have on the results of experiments (Rosenthal 1976; Rosenthal and Rubin 1978). *There is a striking similarity between the experimenter effects discussed within the social sciences and those discussed within quantum mechanics*. The similarity results from the fact that the quantum measurement problem this entire book is based upon is also concerned with the 'influence' an individual might have on the reality it (or better: certain versions of it) experience(s) to what extent. This chapter will argue that the similarity between the two 'observer problems' is *not* accidental. Specifically, it will argue that macro-world experimenter effects in the social sciences might be a *certain form* that the quantum measurement problem takes. This will turn out to have radical consequences for the philosophy of science, i.e., our understanding of the emergence of scientific knowledge, at least for the case of the social sciences this chapter is concerned with, but most probably beyond.

In the next section, some examples for experimenter expectancy effects in the social sciences are reported and discussed. The subsequent section then discusses

© Springer Nature Switzerland AG 2018

C. D. Schade, *Free Will and Consciousness in the Multiverse*,
https://doi.org/10.1007/978-3-030-03583-9_12

how such effects might be integrated into the framework of the clustered-minds multiverse and describes the relationship between experimenter effects in the social sciences and the quantum measurement problem. This section also drafts a thought experiment – quite close to an actual experiment that might be conducted – crystallizing such generalized experimenter effects. The final section briefly concerns itself with the question what are the consequences of those radical thoughts for our understanding of scientific progress in the social sciences.[1] Those thoughts will be continued in the last chapter of this book (Chap. 13), then also with a look at the consequences for research in physics.

Examples of Experimenter Expectancy Effects in the Social Sciences

As had already been reported in one of the footnotes in the last chapter, Rosenthal and Rubin (1978) report on experimenter expectancy effects across 345 studies in the following domains: reaction time, inkblot tests, animal learning, laboratory experiments, psychophysical judgments, learning and ability, person perception and everyday situations. The everyday situations were part of what was analyzed in the last chapter; comparing all other cases, the average effect sizes[2] are largest with *animal learning* (see Table 3 in Rosenthal and Rubin 1978), and the first example I am going to look at is taken from that domain (Rosenthal and Fode 1963). The second example, nicely described in Rosenthal (1976, 139–140) and based upon his unpublished doctoral dissertation (Rosenthal 1956) is stemming from a domain that seems to generally exhibit moderately large effect sizes according to Rosenthal and Rubin (1978): *person perception*. The last example I am going to spend quite some time with stems from more recent research on *social priming* (Doyen et al. 2012).

Let me now start with the especially striking case of *animal learning* and the early experiment on expectancy effects that Rosenthal and Fode (1963) carried out with *albino rats*. Twelve students enrolled in an experimental psychology class served as experimenters.[3] Half of them were told that they are to be carrying out learning experiments in a maze using albino rats and that they were given *maze-dull rats*. The other half was told the same except that they were instructed to be given *maze-bright rats*. The rats were explained to differ for genetic reasons. Each of the experimenters got five rats and had to train them over the course of five days. The reality was that across the two groups the rats were equally bright to start with. But the experimenters that *believed* to have been given maze-bright rats ended up with their rats performing significantly better than the rats of those experimenters that believed to have been given maze-dull rats.

[1]It is not possible within this book to look at experiments outside physics and the social sciences.

[2]For more details on this see the last chapter.

[3]To simplify matters, the role that a thirteenth student played in the experiment by Rosenthal and Fode (1963) will be disregarded, here.

In the discussion section of their paper, Rosenthal and Fode (1963) speculated about many potential explanations for their findings. One explanation that they speculated about (but not endorsed) was even psychokinesis, mentioned also by other researchers that have been trying to make sense of similar results in related experiments: "(…) several workers have even referred, perhaps not entirely facetiously, to the (…) [experimenter's] PK ability"[4] (Rosenthal and Fode 1963, 188). Or in other words, they speculated about whether researchers could be able to *mentally steer* the rats according to their expectations somehow. A couple of other explanations were mentioned in their discussion section such as differences in animal handling: "On a very gross level it might be hypothesized that researchers observing the manner in which a colleague removes a rat from a maze could judge significantly better than chance whether or not that (…) [subject] had performed as (…) [the experimenter] had hoped. (…) An extra pat or two for a good performance, a none-too-gentle toss into the home cage for poor performance (…) may be very revealing to (…) [the subject]. But, it may be said, no "good" researcher would do these things, a point which we may grant" (ibid.). But then, Rosenthal and Fode place an explanation they feel is quite plausible: "While we know little of more subtle cues to animal (…) [subjects], it does not seem farfetched to hypothesize that any (…) [experimenter] may react differently to a well or poorly performing (…) [subject]; and this reaction, mediated by the autonomic nervous system, could well be transmitted to the animal (…) [subject] via changes in skin moisture, temperature, and the like" (ibid.).

Equally exciting with respect to the experimenter expectancy effect shown, but equally vague with respect to the *explanation* of the *workings* of this effect are experimental results on the Freudian defense mechanism of *projection*. In his dissertation, Rosenthal carried out an experiment on human respondents where different individuals (including psychiatric patients) were assigned to three different treatments. The respondents were "receiving success, failure, or neutral experience on a task structured as and simulating a standardized test of intelligence"[5] (Rosenthal 1976, 139). The dependent variable was the rating of the degree of success or failure of persons on photographs. In his *pretest-posttest* design,[6] those ratings were compared between two sets of comparable photographs before and after the treatment. According to the hypothesis of projection, the rated success of individuals on those pictures *after* the treatment would differ from the ratings *before* the treatment in a way consistent with the respondents' experience of their own success manipulated in the experiment. E.g., individuals in the 'success group' should rate the success of individuals on pictures more highly after rather than before the treatment. Since pre- and post-treatment ratings were to be

[4]By doing so, they are referring to a personal communication with Rotter as well as to a paper by Ammons and Ammons (1957).

[5]This implies that—since actual abilities were randomized across groups—some respondents received feedback below, some above their actual level of intelligence.

[6]In a pretest-posttest design, respondents are tested before and after the treatment or experimental condition (with identical or strictly comparable tasks).

compared, statistically, it would have *helped the hypothesis of the experimenter* if the pretest ratings in the 'success group' were comparatively *low* (lower than in the other treatments). And this is exactly what was the case. Indeed, the success ratings of persons on pictures in the *pretest* were significantly smaller in the 'success group' than the success ratings in the other groups. Furthermore, the rating style in this group differed. "[The] (…) success group rated photos significantly (…) less extremely than did the other treatment groups" (Rosenthal 1976, 139). Since the experimenter knew which experimental treatment groups the subjects belonged to, an influence like this would in principle have been possible, but *which way* was the information or influence supposed to travel? "*Whatever the manner in which the experimenter differently treated those subjects* he knew were destined for the success condition, it seemed to affect not only their mean level of rating but their style of rating as well" (Rosenthal 1976, 139-140; italics mine). So the *influence mechanism* is as unclear in this experiment with humans as it is in the above experiment with rats.

Another challenging study of experimenter effects—quite important also for our later in-depth analysis—was carried out in the context of a *social priming experiment*. Doyen et al. (2012) build up upon the well-known paradigm by Bargh et al. (1996) "in which participants unwittingly exposed to the stereotype of age walked slower when exiting the laboratory" (Doyen et al. 2012, 1). Doyen et al. not only criticized and modified the original experiment but also analyzed it with respect to the impact of experimenter expectancy effects. Let me go into some detail, here. The original experiment by Bargh et al. (1996) "involved asking participants to indicate which word was the odd one out amongst an ensemble of scrambled words a number of which, when rearranged, form a sentence. Unbeknownst to the participants, the word left out of the sentence was systematically related to the concept of "being old." (…) Those participants who had been exposed to words related to old age walked slower when exiting the laboratory than the participants who had not been so exposed" (Doyen et al. 2012, 1). However, Doyen et al. (2012) were skeptical with the results because of *possible experimenter expectancy effects* as well as the use of a *manual stopwatch* in the experiments by Bargh et al. (1996).

In the first of two experiments by Doyen et al. (2012), two infrared sensors were hidden in a hallway that participants went along after the experiment. Whilst keeping the measurement distance the same as in Bargh et al. (1996), this methodological change, according to the terminology of the authors, made the measurement of walking speed *objective*. Again, there was a prime and a no-prime condition, the prior exposing respondents unconsciously to the concept of old age. "Each experimenter randomly tested participants from both conditions and was instructed to interact with each participant according to a strict script so that their potential influence was minimized. (…) The questionnaires were enclosed in an envelope that the participant had to open, so as to keep each experimenter blind of the participant's condition"[7] (Doyen et al. 2012). Using this setup, Doyen et al. (2012) were not able to replicate Bargh et al.'s (1996) findings. There was no significant difference in walking speed between the primed and the non-primed groups.

In their second experiment, Doyen et al. (2012) manipulated "(…) experimenters' expectations about primed participants' behavior (…). One half of the experimenters were told that the primed participant would walk slower as the result of the prime (…), the other half were told that the participants would walk faster" (3). Walking speed was measured using both the infrared gate as well as a manual stopwatch. The results were striking. Looking at the manual stopwatch results, experimenters *fully* confirmed expectations based on their manipulation, i.e., reported *slower* walking when so manipulated with the primed group, and even *faster* walking with this group when manipulated that way. Sure enough, measurement errors possible with the mechanical stopwatch were (unconsciously) '*used*' by the experimenters to make results confirm their expectations. No ambiguity remains with this part of the results.

The situation becomes far more interesting when looking at the *objective measurements* using the infrared gate. In this case, the no-prime and the prime group did *not* differ in walking speed when experimenters were manipulated to expect *faster* walking after having been primed on stereotypes of old age. However, there was an objective decrease in walking speed, confirming the result by Bargh et al. (1996), in the treatment with experimenters' expectations manipulated towards expecting *slower* walking of respondents being primed with stereotypes of old age. The interpretation by Doyen et al. (2012) is hard to follow. Let me therefore try to reorganize it a bit, disregarding the results achieved with the manual stopwatch, but looking at both experiments together. There are three cases, depicted in Table 12.1.

Are the findings consistent with some straightforward psychological theory? Apparently not. Firstly, as in the two other examples analyzed in this section, the authors finally rely on the vague notion of *environmental cues* and *experimenters' behavior* to explain the experimenter expectancy effects they receive. It is unclear what kinds of cues these are supposed to be. Secondly, environmental cues and experimenters' behavior are not strong enough to *override* some (nevertheless!) *basic effect of behavioral priming* somehow; there is no conclusive evidence for a pure self-fulfilling prophecy effect. But then, thirdly, when nothing is expected (?), behavioral priming is absent? I would like to argue that the interpretation of the findings, especially the explanation of the emergence of experimenter expectancy effects, is quite unsatisfactory in the study by Doyen et al. (2012): An interaction effect between unknown cues with an unknown transfer mechanism and behavioral priming, but only when this is set up in the right direction? This explanation is as

[7]Other features of the experimental design, e.g., the indirect measurement of respondent's awareness of the priming, are disregarded, here. In fact, respondents showed some degree of awareness of the prime.

Table 12.1 A reorganization of the 'objective' priming findings by Doyen et al. (2012)

Experiment/treatment	Experimenter's expectancy	Results	Interpretation by Doyen et al. (2012)
Experiment 1: primed respondents but experimenters not influenced	None? (But who is the 'experimenter' then, the primary investigators?)	Speed does not differ from control group	Shows that there is no general effect of behavioral priming
Experiment 2: primed respondents and experimenters' expectations influenced in line with the theory by Bargh et al.	Slower walking speed expected	Respondents slower than in control group	"primes (…) must be in line with environmental cues such as the experimenters' behavior" (Doyen et al. 2012, 6) to get the result
Experiment 2: primed respondents and experimenters' expectations influenced opposite to the theory by Bargh et al.	Faster walking speed expected	Speed does not differ from control group	"results (…) cannot be explained solely in terms of (…) self-fulfilling prophecy effect" (Doyen et al. 2012, 6)

unsatisfactory (or even more unsatisfactory) as the explanation of the findings from the much older study by Rosenthal and Fode (1963) and from Rosenthal's dissertation discussed before.

I am going to come back to the results by Doyen et al. (2012) in the next section, when the more general question is analyzed as to whether we are looking at the right *type* of explanations for experimenter expectancy effects, whether, e.g., 'subtle cues' such as skin moisture (Rosenthal and Fode 1963), mentioned in *all* three above studies are the full story, here. Doyen et al. are certainly aware of the fact that their explanation of expectancy effects is not conclusive: "Experimenters' expectations seem to provide a favorable context to the behavioral expression of the prime. Obviously, this interpretation remains tentative, *as we do not know how this process operates*" (Doyen et al. 2012, 6; italics mine). The progress that has been made in this regard since the 1950s and 1960s appears to be somewhat limited.

Experimenter-Expectancy Effects as Measurement Effects in the Clustered-Minds Multiverse

Has the literature been telling the wrong story of the emergence of experimenter expectancy effects so far, or has that story just been incomplete? What is the correct and/or complete story, then? As was the case with outside-laboratory self-fulfilling prophecies in Chap. 11, the 'story' I would like to suggest here is one where the researcher expects certain findings from his experimental study and allocates, consciously or unconsciously, a high amount of consciousness to those realities containing preferred experimental outcomes. So, does this imply that I want to *reject*, for instance, the 'subtle cues' explanation—suggested by the respective authors as an explanation for the results of the experiment with albino rats (Rosenthal and Fode 1963)? Do I want to *replace* this explanation with the clustered-minds-multiverse explanation? Am I even coming back to something like the psychokinesis explanation discussed by those authors? The 'story' is not as simple as either the first or the second suggestion might imply, and it again shows how different some of the interpretations via the clustered-minds multiverse are from more traditional scientific explanations.

Firstly, allocating a lot of consciousness to those 'movies' where our *experimenter expectations come true* might simply imply putting more emphasis on 'movies' where cues such as facial moisture *work* and it implies putting less emphasis on 'movies' where they do not work. This might work 'ex ante,' i.e., in form of a vectorial choice, and/or 'ex post,' in the form of consciousness reallocations. (Avoiding too extreme allocations of consciousness would, moreover, probably lead to both results exactly in tune and results halfway in tune with the experimenter's expectations occurring in different high-consciousness realities.) But even this might still be too simple a view: If it is not the facial cues, the relevant 'movies' might 'find' different 'ways,' as long as they lead to the expected result—perhaps even psychokinesis or twitchy legs, for that matter. Thus, the fact that researchers will sometimes be finding straightforward explanations for experimenter expectancy effects—establishing a transparent causality with known factors—and sometimes not, is *not meaningful* for the *fact* that the measurement result is only one out of many occurring with different versions of the individual (and realized in different realities).

Secondly, being able to put more 'emphasis' on those realities where our expectations turn out to be fulfilled also requires one to not cross the boundary to a very-low-consciousness situation. In social science experiments, this requires 'finding' or 'convincing' enough versions of different individuals to allocate more than a minimal[8] amount of their consciousness to realities where they are behaving in a way that is consistent with the researchers' expectations (e.g., in the case of more of less maze-dull rats, being attentive enough to read the cues in the experimenter's face).[9]

Thirdly, each result that is published and read is published and read within a certain *minds cluster* (see also Chap. 13). This implies that there finally must be people sharing a reality with the experimenter where his results are believed or at least be taken seriously enough to be scientifically processed and discussed. At the same time, there will be other versions of results for the same research question that occurred either with different individuals or with different versions of the same

[8]As usual, if only very-low-consciousness versions of the respective individuals are available, the researcher has to avoid that situation.

[9]A discussion as to whether or not rats need to possess a higher degree of consciousness to behave in the way supposed here and whether they actually do possess this level of consciousness is beyond the scope of the book. If you do believe that they would need to possess a higher degree of consciousness for the experimental results to occur and that rats do not possess it, please replace the rats (and the respective experiment) by people (and a different experiment). Then you are, however, back to the question how to explain the result by Rosenthal and Fode (1963).

individuals that are published and read in a different minds cluster. The good news is, however, that we (you, that version of the reader reading this version of the book, I, that version of mine writing this version of the book, as well as many 'others' that share a minds cluster with the two of us) will be confronted with at least halfway consistent scientific results within that minds cluster.

Is this mere speculation, or could such generalized experimenter expectancy effects or *generalized Rosenthal effects*—in the sense of a vectorial choice with respect to experimental results, irrespective of the path of realization, measurable or unmeasurable—actually be shown? The answer to this question can already be discussed based on existing experimental results. Turning back to the paper by Doyen et al. (2012), one interesting aspect was the *null* result in their first experiment (i.e., their baseline treatment) without any manipulation of experimenters' expectations and just replicating—with some technological and design improvements—the experiment by Bargh et al. (1996) matching the *null* result by those experimenters in the second experiment manipulated to expect the opposite effect to that found by Bargh et al. (1996)—individuals walking faster after being exposed to the old age prime. A multiverse-interpretation consistent with those results would be that the *principal investigators*, i.e., Doyen, Klein, Pichon and Cleeremans, *expected behavioral priming not to work* (that would not be surprising, given the setup of their study and their line of arguments). Of all available realities where behavioral priming, say, turns out to have an impact on people's behavior (let me suppose that behavioral priming, as one of the 'other factors,' indeed works on average),[10] the effect ranging from hardly above zero to large, they put most conscious emphasis on that reality with the smallest possible effect size.

In the second experiment, the manipulation of experimenters' expectations opposite to the theory of Bargh et al. (1996) might have again emphasized that reality where the basic effect of the 'other factors' (the basic priming effect) is countered, but in a different way. One could have expected those effects (negative expectations of the principal investigators) plus manipulation of experimenters to expect the opposite of the 'other factors,' in the form of social priming effects, to add up. This did not happen, perhaps because, in the eyes of the principal investigators, the manipulated experimenters just took over their 'role' as critical evaluators of the theory of social priming, somehow, thus bringing in nothing 'additional' to strengthen the phenomenon.

Is there a way of substantiating such thoughts even more? Let me try to create a thought experiment that might in fact, with some alterations, be turned into an actual experiment; an experiment that would directly test for experimenter expectancy effects in the sense of the quantum measurement problem, excluding known ways of information transfer from the principal investigators to the respondents as

[10]Some would debate that (however, I am not one of them). But it is interesting how far I can actually get with my argument assuming that the basic social priming effect exists.

much as possible.[11] To do so, let me look at one more game created by Kahneman and coauthors, the *dictator game* (Kahneman et al. 1986), and make this the basis of a meta thought experiment. In the basic dictator game, one of two players (the dictator) gets some amount of real cash to start with, an endowment of, say, $20.00. He is then randomly (and typically anonymously) paired with one other player. The dictator knows that the other player received no endowment, the other player knows that the dictator received $20.00. The dictator is now asked what amount of money he is willing to transfer to the other player. He is allowed to choose any amount between $0.00 and $20.00.[12]

The dictator game has been suggested by Kahneman et al. (1986), mainly to test for *fairness*. Others have suggested that the behavior of many respondents in dictator experiments might predominantly reflect *altruism* (Eckel and Grossmann 1996). But this is not such an important distinction for my analysis, here. The dictator game is important and controversial for a different reason. Whereas some researchers, i.e., standard *rational choice* economists, believe in a *purely self-serving homo economicus*, most *behavioral economists* or economic psychologists (including Kahneman et al. 1986) would argue that individuals' motives contain fairness and/or altruism. Thus, the expectations of rational choice researchers and behavioral economists with respect to behavior in this game would be dramatically different.[13] Let me assume that we are in the early 1980s, and no such game has been carried out so far. Whereas the rational choice economists would now predict that individuals would give *nothing or very few* in this game (perhaps just enough not to feel uncomfortable for being perceived as stingy), a typical behavioral economist would predict that the dictator will be giving *substantial amounts* to the other player.

In my thought experiment, I will now let a rational choice economist run the experiment with n = 200 respondents in one city. At the same time, a behavioral economist believing in altruism or fairness would be running the experiment in another city (where the respondents do not differ in any relevant aspect from the respondents in the other city).[14] He also runs the experiment with n = 200

[11]As already mentioned, the fact that normal means of information transfer would be identifiable in the experiment would not reject the multiverse interpretation of the experimenter expectancy problem. However, many would then argue that this theory is *not falsifiable*. Therefore, it would be advisable for this experiment to be devoid as much as possible of any ways of interpreting the effects via traditional mechanisms.

[12]The difference between the dictator and the ultimatum game dealt with in chap. 10 (in the section on interaction with humans and computers) is that in the dictator game, the other player has no possibility to reject the split proposed by the player who owns the endowment.

[13]Whilst an overview of the numerous studies that have been carried out since 1986 is not possible in this chapter, the interested reader might be directed to a meta study on this matter by Engel (2010).

[14]In actual experiments, two different cities would not suffice as a solution, at least not without further design features. Individuals in two different cities might differ on many dimensions, including dimensions that have an impact on the experimental treatment variable (here: on altruism and fairness).

respondents. Let me furthermore assume a sophisticated experimental design by both; a design that excludes regular experimenter effects, mainly by letting *other individuals* (experimenters) run the experiment that get no information whatsoever on theory and expectations of the respective principal investigator, perhaps by getting only written, standardized information.[15]

Let me now assume that the following set of results is observed (in our minds cluster; thus, to be sure, in one version of reality, only). The monetary amounts transferred in the experiment planned by the rational choice economist are very low, only 5% on average, with many dictators not giving anything to the other player. Whereas the monetary amounts transferred by dictators in the experiment planned by the researcher believing in fairness and altruism would be much larger with an average transfer of 35% of the money, with a considerable portion of individuals suggesting an equal split and with only very few dictators not giving anything.[16]

Firstly, given the purpose of my thought experiment, this finding would underline the existence of a *generalized Rosenthal effect* because the difference between the results of the two researchers is substantial, and because the only *systematic* difference between the two experimental setups, by definition, was the different expectations of the two principle investigators that were not even in contact with the respondents and that did not reveal their expectations in any way to the experimenters running the experiments. Clearly, this view is a simplification, since there is no *direct* causal relationship between experimenters' expectations and the outcome achieved in our minds cluster. Indeed, what we observe *here* is one of many realities. Secondly, a *mechanism* of the functioning of this experiment might become evident, a mechanism close to the thought experiment by Wigner in quantum mechanics, *Wigner's friend* (Wigner 1983). I have mentioned the *usage of non-biased experimenters* that help each of the two researchers to carry out the experiment. Indeed, this is the typical *cure* suggested to get rid of normal Rosenthal effects: "Whenever possible, the actual running of (…) [subjects] should be done by research assistants who do not know what outcome is desired" (Rosenthal and Fode 1963, 188). In the thought experiment suggested by Eugene Wigner to explain the central role of consciousness in the measurement problem of quantum mechanics, not the principal researcher but an assistant is conducting measurements at some quantum system. In *his* thought experiment, the question is *when* the measurement is actually conducted: when the assistant has looked at the measurement device, or rather when the principal researcher has learned the result from his assistant (Wigner 1983)?

[15]Let me furthermore assume that both principal investigators use identical experimental instructions (perhaps based on some prior exposition of the experiment as a hypothetical one in some publication) so that the difference cannot be explained based on that difference.

[16]Note that there might be realities where each of those experimental results turns out to be even more extreme, i.e., more biased towards the expectations of the respective researcher; however, the two researchers would then most likely not *both* be present in those realities with sufficient consciousness.

In a consequent application of the clustered-minds multiverse to *my* thought experiment based on the dictator game, it would make sense to seriously consider the possibility that it is the *consciousness of the two principal researchers* (and not the consciousness of their assistants) that is decisive for the experimental results achieved; since it is the principal researchers that will analyze the findings, write up the results and publish the findings. It is *their view on reality* that should be dominant. The consequence of this thought, however, is quite radical: It would not help much to employ assistants to run the experiment. Whereas this might help getting rid of some of the 'regular' parts of experimenter expectancy effects, those parts that, frankly, *appear* regular; but there might be *no* possibility of getting get rid of the entire, generalized Rosenthal effect. (Note that in my interpretation of the findings by Doyen et al. (2012) I have been slightly less radical and allowed for some impact of the experimenters' consciousness, too.)

Consequences for the Progress of Scientific Knowledge in the Social Sciences

According to *critical rationalism* (e.g., Popper 2014), one of the most well-known epistemological philosophies and continuously part of most monographs and courses on the philosophy of science since shortly after its inception, scientific research is supposed to make use of a specific combination of *induction* and *deduction* to achieve scientific knowledge.[17] Specifically, the researcher is supposed to start with a scientific hypothesis (based on a theory, ideally) and then to test this hypothesis empirically. In fact, the researcher is supposed to try to *falsify* it. If the hypothesis is falsified, it has to be modified. If that hypothesis was based on a theory, the theory must also be rejected (or modified). A majority of empirical scientists from different fields implicitly or explicitly organize their research according to the principles of critical rationalism. This is especially true in large parts of the social sciences.

But what happens if one *uses* the framework of the clustered-minds multiverse to reanalyze this thinking? Does, e.g., having a hypothesis and/or a theory make the researcher less 'neutral' with respect to the outcomes of an empirical study? The researcher is supposed to try to reject his theory, but this is not too plausible, psychologically. I would rather expect him to try the opposite, sometimes perhaps unconsciously, and to allocate a lot of consciousness to a reality where the theory is confirmed. At least this is what the examples and the reasoning within this chapter have been about. However, no matter whether one or the other approach is correct

[17]A thorough discussion of different epistemological concepts and different approaches to the philosophy of science are beyond the scope of this book. But see the critical discussion of Adlam's (2014) approach, partially based on Deutsch's (2016) thoughts within Chap. 2. According to Deutsch (2016), a research program following critical rationalism helps making Everettian quantum mechanics be testable (Deutsch 2016) (see Chap. 2).

as an empirical description of the behavior of researchers, verification or falsification, the researcher *will* have, in the form of a vectorial choice or consciousness reallocations, an influence on how much conscious emphasis is put on one form of the result or on another form of the result. So, the idea of coming up with scientific hypotheses and trying falsification (or verification) might not have an unambiguously positive effect on the progress of scientific knowledge. The sophisticated logic of critical rationalism might have slightly problematic consequences if we are indeed operating in a the clustered-minds multiverse.[18]

But let me now suppose that we do *not* start our scientific investigations with theories and hypotheses. Is this a solution? Certainly not. First of all, there is the entire *problem of induction*. Here is a well-known example: If one has collected empirical knowledge about the color of swans and came up with *150,311 swans that are white*, does this justify a theory saying that all swans are white? Indeed, having no (e.g., biological) theory the color white has been deduced from and that has survived many falsification trials leaves us quite unprotected against the 150,312th measurement result being a *black swan*.[19] But without going deeper into this well-known discussion, here, would this, secondly, even be helpful with the basic problem of expectations *directly* driving the conscious emphasis that is put on different scientific results? I guess the question I have to answer here is whether or not one can ever be free of (unconscious) expectations, even without possessing any theory and having no hypotheses in mind? The answer might be "No." But if this is impossible, having an *explicit* hypothesis makes our expectations at least transparent —in many cases perhaps expecting just the opposite of falsification. In any case, the consequence of looking at critical rationalism and scientific progress from the perspective of the clustered-minds multiverse is that *falsifications* might be *less frequent* than they would be in a singular universe. If falsifications are rare, *many* theories survive. Is it then possible that many theories are *true*, even some that (partially or fully) contradict each other? My answer would be a tentative "Yes," but not necessarily within one reality; rather if they hold within *different minds clusters*.

Let me now return to the thought experiment based on the dictator game. One researcher finds individuals to be 'giving,' the other finds them to be extremely cheap. Let me furthermore suppose that other researchers—with varying expectations—try to replicate the experiment. We are used to contradictory findings in scientific research, but perhaps the findings within *one group* of researchers are somewhat similar, at least quite consistent with each other; but considering the results by *some other group* of researchers, they look very different from those in the first group. Again: Will all those findings appear in one reality cluster? Perhaps to some extent, but this is really unclear. There might even be minds clusters where

[18]This might appear as a contradiction to Deutsch's (2016) argument that critical rationalism might be used to craft (or ex-post justify) an empirical strategy to test Everettian quantum mechanics (see Chap. 2). But it is not. It just implies—and Deutsch would most certainly agree—that the conditions for theory testing are simply harder in the multiverse for several reasons.

[19]Pardon the usage of this example; it is very illustrative, but I do not want to evoke any connotation with Taleb's (2007) analysis, mostly concerned with randomness and rare events.

people in general *are* very unfair, and where nobody is surprised that a majority of publications reports on very small amounts of money given by the dictator to the other receiver. There might be other minds clusters, however, where people *are* much more generous, and where transfers in the dictator game are considerably larger. It is hard to say which set of findings is 'correct' or 'more correct' or which reality produces 'more objective' results. The question of an *objective truth* might even be seen as ill posed.

It is clear, however, that replications become more important within the clustered-minds multiverse than has already been conjectured in many fields of science within a singular-reality worldview. They at least offer a chance of getting closer to something that might be called scientific progress because they might give an idea of the scope (or distribution) of possible outcomes of a certain type of experiment (albeit even the distribution of results might differ between reality clusters); they do not, in any case, protect from the *subjectivity* or *cluster-specificity* of knowledge. This topic (among others) is analyzed in more detail in the next (last) chapter of the book. Even though times are just special cases of parallel realities (see Chap. 3), there might at least be two dimensions that organize our understanding of the sorting of versions of individuals/realties into clusters, relevant for the advancement of scientific knowledge:

(a) Different knowledge clusters that exist at the same point in time.
(b) Knowledge clusters that exist at different points in time.

Some clusters within both cases might naturally be called *paradigms* in the sense of Kuhn 1996 [1962] because of knowledge changes (or different views on scientific truth, really) that ask a lot of people that are supposed to mentally 'move,' in the sense of either re-allocating consciousness to versions of theirs residing in different minds clusters[20] [in case (a)] or 'organically' developing, along one 'decoherent history,' together with the rest of the cluster (or some part of the cluster), to a new paradigm [in case (b)].[21]

[20]It is unclear to me as to how frequent the case of individuals' versions residing within different minds clusters is; I am currently not able to think of any good reasons as to why this case should be rare. However, moves like this, if they are substantial enough, require the application of 'quantum brainwash,' as has been pointed out in Box 6.3.

[21]In some of those cases, the difference between vectorial choices and consciousness reallocations might become blurred.

PART V
CONCLUSIONS AND GENERAL PERSPECTIVES

Chapter 13
Selected Consequences of the Clustered-Minds Multiverse for Weltanschauung and Scientific Research

This Book Is One Possible Version of the Book, and Many Others Exist

The reader who has followed me up to this point of the book should not be surprised when I am now saying that he or she is sharing a reality with me as well as with the other readers that have been reading or will be reading the book in *this* version and not in any other. *We are in a minds cluster.* And that there are other reality clusters where other versions of mine have been (or will be) writing this book in a different version. Some versions of the book might be more radical, some less radical, some are longer and some are shorter. With some versions, the author has been younger, with other versions he will be older. In some he is using more mathematics etc. Some versions of mine might be too occupied with other things throughout their lifetime to ever write it.

Therefore, it should be clear, when I am now talking about weltanschauung and future research opportunities, that I am not claiming to talk about *the* weltanschauung, *the* scientific truth and *the* further development of it, but most probably only one of many truths, a truth that (hopefully) holds within our (probably large) minds cluster. (And that will naturally be debated within it, too.) Whereas I would like to stress that the *fact* that we are living in a multiverse is fairly robust and the clustered-minds multiverse a plausible version of it, some of the consequences that I have derived from it have most certainly been derived differently in other existing versions of the book (i.e., in parallel realities).

Does the latter sound trivial? Already in a singular reality, opinions would differ, scientists would argue about appropriate theories, about the correct interpretation of experimental results. But the situation is *more extreme* in the clustered-minds multiverse. Because there *are* variants of this book, but also variants of physics, psychology, or economics books, variants of certain philosopher's interpretation of quantum mechanics etc., even though *we* cannot find, buy or read them. I think it is important to keep that in mind when reading this last chapter of the book

© Springer Nature Switzerland AG 2018

C. D. Schade, *Free Will and Consciousness in the Multiverse*,
https://doi.org/10.1007/978-3-030-03583-9_13

(and reconsidering what has been encountered in the other chapters, perhaps). This last chapter will concentrate on two subjects: (1) It will refer back to those results from the different chapters that have a special importance for our weltanschauung as well as to those that have a special importance for the title issues of free will and consciousness—what is the clustered-minds-multiverse worldview (in a nutshell), and what does it have to say on free will, consciousness and, more generally, on decision making? (2) It will discuss some selected consequences for future research, it will mainly look at general consequences for the generation of scientific knowledge, based on the multiverse perspective, and at some types of studies that might follow from (or be encouraged by) this book. Note that an overview of the plethora of research opportunities revealed throughout the chapters of this book is impossible as well as unnecessary here—the latter because the chapters were quite explicit about them. Those two subjects in turn determine the two main sections of this chapter.

The Clustered-Minds-Multiverse Worldview

Newtonian, Quasi-Newtonian and Multiverse Worldviews

The first chapter of this book started with Freud's idea of the *narcissistic wounds* that scientific research and thinking inflicted on mankind, with the heliocentric worldview and the detection of the unconscious as the most prominent ones. And I have there already started discussing whether the *acceptance of the existence of a multiverse* with all its consequences such as the existence of numerous versions of ours may or may not cause another wound. In fact, I have argued that the multiverse does not only imply that we are *not unique* but only one of many versions of ours, perhaps a *potential* wound, but that it has *many positive aspects to offer*, that it might in fact *heal some important other wounds*, countering the claims that we have no influence on what we experience in our lives or that consciousness is only an epiphenomenon of physical processes, that it supervenes on the physical.

Especially in the social sciences, most would nowadays still implicitly equate a physicalist' worldview with a Newtonian one implying a mechanistic world that works like a large and complex clock. Since the Newtonian worldview makes accurate predictions with large objects in the type of 'reality' we are used to (perhaps given the alterations by special or general relativity theory that seem not to bother us much in our everyday life), it is not surprising that quantum mechanics—that seems to turn that world on its head—was first interpreted in a way that seemed to allow staying in our comfort zone. The standard or collapse interpretation (or Copenhagen interpretation, in the sense of an umbrella term), assuming a *collapse* of the wave packet, allowed us to keep something I would like to call a quasi-Newtonian worldview. This was correct even when it turned out to be impossible to maintain the artificial 'cut' between micro and macro world. Indeed, even the more sophisticated (and more recent) explanation of the appearance of the macro world, *decoherence*, has been applied in a way that could somehow *be made consistent* with a quasi-Newtonian worldview (and has been by many). A longer section in Chap. 2 was required to demonstrate that the mainstream usage of decoherence is again containing some important hidden but unjustified assumptions, perhaps with the *aim* of maintaining a quasi-Newtonian worldview.

The big contribution by Everett (1957) (consistent with a different take on decoherence[1]) was to free quantum mechanics from the ballast it was carrying; the ballast of people enforcing the appearance of something that might not exist: a singular, quasi-Newtonian or quasi-classical reality. But Everett's thoughts were then either ignored or, left with the 'quagmire' of the wave function, mostly absurd interpretations of the resulting multiverse were suggested. Interpretations that were absurd enough to *discourage* people trying to work with the multiverse interpretation and still not radical enough by keeping, e.g., the problematic assumptions regarding the preferred basis those multiverse interpretations shared with the standard interpretation. Freeing the multiverse interpretation from some of the absurdities on the one hand and adding some plausible restrictions on the other hand (e.g., an uneven, but not binary distribution of consciousness among different realities) led me to a *new weltanschauung* that might be considered breathtaking in two regards: (a) Reality is largely *constructed* by consciousness, under the restrictions of the wave function, entanglement, and the Born (1926) rule as an auxiliary equation for quantum measurements along one decoherent history (see Chap. 4), with different experienced realities (or 'movies') being partially the result of *collective decisions*, i.e., decisions within minds clusters; and (b) consciousness is (under some restrictions) able to freely implement vectorial decisions, i.e., what realities to experience with what 'intensity'[2] (with which version of the individual).

What Has Been Learned for Free Will and Consciousness?

Free will exists, albeit not in an unlimited form, and it furthermore works in a way that differs from what most might intuitively think how it works. It is limited, e.g., by the distribution of others' consciousness, i.e., by the need to avoid very-low-consciousness situations, as well as by the requirement of choosing moderate allocations of consciousness. It is also limited by our own unconscious motives, that might have an effect similar to conscious free will but without us realizing those motives (revisit the list of factors impacting of vectorial choices, potentially limiting free will, presented in Table 8.4 and still being incomplete).

[1]Of course, Everett (1957) preceded decoherence (Zeh 1970). But applying decoherence within the Everett framework leads to a more parsimonious model of the appearance of a seemingly classical world.

[2]Simplifying, the intensity is here equated with the amount of consciousness allocated to some version of the individual.

Meaningful clustering, although only vaguely understood at best (see Chap. 10), might finally have a limiting effect on free will, too. Free will means putting different emphasis on different 'movies,' and the allocation is allowed *not* to be binary; but it is also *not* allowed to be binary. Or in other words, consciousness in some reality can never be equal to zero, and, consistent with that, one can never allocate the entire consciousness to one reality; a 'normal' case might be (we do not actually know) that we put a lot of emphasis on a moderate number of realities, and that the rest of consciousness is then distributed (perhaps unevenly) among the remaining ones.

Let me finally note that reallocations of consciousness are necessary components of vectorial choices and hence of free will, even though one might in general not realize this action and it leads to the slightly odd construct of an unconsciously executed kind of free will, at least on the level of the individual's version we experience; or in other words, this type of free will pertains to the individual, but not to its versions, somehow. Let me remind the reader of the fact that free will, however, executed within vectorial choices, cannot be experienced, in a narrow sense, either, because we only experience one version making one choice. Thus, the difference between free will executed within vectorial choices and within reallocations of consciousness is not that huge, after all. Unfortunately, I will not be able to pursue this matter further in this book. Let me also remind the reader of the importance of reallocations of consciousness because otherwise, consciousness might be diluted more and more with each branching. However, since reallocations of consciousness might also be driven by many of the factors listed in Table 8.4, I suppose, I have not been dealing much with them at their own right (the most important application has been that reallocations might partially counter the real-options structure of many vectorial choices/reality selections in the multiverse).

According to the concept of dualistic idealism (proposed in Chap. 5), we are not living in a world of 'brick and mortar,' and we are not part of a large clockwork. Physicalism, in the narrow sense of sort of a singular-world materialism often ascribed to it, is untrue. Instead, following the multiverse interpretation developed in this book, reality loses its 'substance,' somehow, it is rather illusionary (Maya). But that in turn provides the basis for an only loosely restricted freedom to experience what we prefer (to the largest extent this applies collectively, within a minds cluster). Thus, as already mentioned, whereas some individuals might perhaps experience it as another narcissistic wound that they are *only* one out of several versions of theirs, and that they are able to experience only a small part of a larger reality, that our life is a journey of consciousness meeting with other consciousness within a wave function rather than a way through actual underwood or a hike through actual mountains, others might put more weight on the fact that the next wound that was waiting for mankind or perhaps already started to hurt: *neurobiological reductionism* or the dissolution of the dualism of body and soul (see Chap. 1), can be avoided, or *healed*, respectively, by adopting the multiverse perspective developed in this book.

And this is important for various reasons. First of all, psychological issues are plausible to arise if one really understands that many researchers, deriving their thoughts on the basis of a deterministic, singular universe, want to imply that free will is an illusion or even consciousness—the center piece of how we make experiences—is. Secondly, within the clustered-minds multiverse, (partial) responsibility exists. Whilst some authors, so-called compatibilists, try to save the notion of responsibility even without any possibility to choose otherwise (i.e., in a deterministic, singular universe), I have argued that such an account is not very convincing (see Chap. 7). On the other hand, the case for responsibility is not quite 'clean' in the multiverse, either. Looking at singular decisions (i.e., without explicitly considering future or past decisions), responsibility is clearly debatable within vectorial choice; it might sometimes be hard to impossible to completely avoid 'bad,' unlawful realities even if the core of consciousness resides with a 'good,' highly ethical course of action. However, with multiple choices and in a long-term perspective, holding an individual responsible for his actions and personal *development* is rather justified (see Chap. 7).

One important implication of the clustered-minds multiverse and free will has only received limited attention so far (in Box 6.2 in Chap. 6). This aspect is the result of the *preferred basis problem* that was briefly analyzed in Chap. 2. To quickly repeat, most interpretations of quantum mechanics—including most multiverse interpretations that have so far been suggested—assume that we are naturally experiencing walls, dogs, houses and cars. In the terminology of quantum mechanics, this is, as has been explained, already, a matter of the preferred basis of the state space that is mostly assumed to be chosen in a way that 'pure' or 'natural' measurement results appear. The superpositions are then composed of such 'natural' alternatives as Schrödinger's alive cat and Schrödinger's dead cat. It is hard enough to swallow that the cat might be alive *and* dead in different parallel realities, but at least the measurement will always lead to one or the other state of the cat appearing in our consciousness. If the preferred basis problem is taken seriously, however, not even the alive cat or the dead cat might be the basis for the superposition, but already mixtures of the two or even dog-cats.[3]

This book followed the lead by Lockwood (1996) in suggesting a radically subjective solution of the preferred basis problem in Chap. 2. Page (1995) stated that this additional multiplicity of possibilities due to non-classical worlds leads to a situation of *many-many-worlds*. But what is the consequence of this matter for the discussion on free will and consciousness, what is the issue we are here facing? The consequence of many-many-worlds is that there is a second form of free will, however a potentially *dangerous* one. This second form of free will is somewhat close to *craziness* in violating perceptual conventions within a minds cluster (at least within my minds cluster and hence, since I am sharing it with my readers: yours). Perceiving dog-cats is not the *norm*. Moreover, it might imply violating a restriction (again, at least in our minds cluster) I have posed early on and then maintained throughout the remaining

[3]This example is similar to the one in Schade-Strohm (2017), patch 1: "A different history of creation," where a 'little entity,' not proficient so far in creating objects in consciousness, creates a sheep-dog.

book: not entering very-low-consciousness situations. Who are you expecting to meet when you are perceiving (and, say, talking about) dog-cats? But we all know that some individuals *have* mental problems. This is perhaps a novel way to see and treat the problems they have. The mere fact that at least some of those individuals can be told that they are leaving perceptual conventions and, possibly, nothing more 'serious' can be detected in their brains etc., might already be helpful for them. It might 'de-pathologize' their symptoms. But there might be even more to it. Patients suffering from schizophrenia might have to be understood as valuable contributors to a better understanding of the universe if they are considered to be people with a better access to parallel realities than other individuals.[4]

What Are the Consequences for Our Decision Making?

Free will and consciousness are important philosophical problems, and they have—in one form or another—occupied a major part of the book. However, there is also a 'practical' aspect to the type of free will we possess and the role that consciousness plays in the clustered-minds multiverse: that there is a certain way how we actually make choices. I have dealt with this issue mainly in Chaps. 6, 8, 9 and 10. Specifically, I have tried to shed light on the question as to how the decision sciences might have to be altered to take into account the framework offered by the clustered-minds multiverse, including the important question how free will is executed. This required to introduce a new theoretical concept: vectorial choice. I have to admit that many of the presented thoughts in Chaps. 8, 9 and 10 were somewhat speculative. But I feel that a number of research opportunities are arising in this domain and that the discussion was therefore fruitful.

Macro-world probabilities have a *subjective meaning* in the sense of a 'lack of knowledge' of future developments relevant for different versions of the individual. Also, there are conscious and unconscious influences an individual's consciousness exerts on the amounts of consciousness allocated to different realities in vectorial choices. Many decisions in the multiverse exhibit a real-options structure. Given the complexity of decisions in the clustered-minds multiverse, the effectuation principle (and further developments based on it) has been suggested as an appropriate tool to handle them; this at least being a temporary fix, perhaps more, the hope still being that a normative theory of decision making can be developed for vectorial choice.

[4]Similar considerations are perhaps valuable for a better understanding of dreams. Freud considered dreams as being generated by the unconscious of the individual, and this is certainly true to some extent (Freud 2010 [1900]). A complimentary perspective, however, is that dreams might offer an access to parallel realities (similar thoughts can be found within different chapters in Mensky 2010).

A feasible notion of utility—as a guideline for individuals' vectorial choice—turned out to be, anyway, hard to define. The problem already starts outside the multiverse, i.e., with regular choice of singular alternatives. Within the multiverse, the situation proved to be especially difficult (see Chap. 9). And according to Chap. 11's example of repetition compulsion, it is all but clear what the unconscious goals are that people follow within their choices; I also introduced the concept of utility interdependence between different versions of an individual and discussed potential ways how this might be 'set up' in the multiverse (see Chap. 9).

Chapter 10 on strategic and economic decisions led to twofold consequences: research opportunities within game theory and economics *applying* the clustered-minds multiverse, but also results from simultaneous market entry experiments *adding* some ideas to the understanding of how clusters of minds are actually formed (in the sense of a 'meaningful clustering;' see Chap. 10).

I do not think that people are equipped, at this point, with a *multiverse toolbox* to make better choices. At least the knowledge offered in this book is not 'plug-and-play.' But individuals might be, however, encouraged to reflect upon their decision making in various novel ways; and this in turn might eventually be valuable for their choices. Intensive work as well as a long-term research perspective are needed to elaborate on the laws and success factors of decision making in the clustered-minds multiverse.

Future Research: How the Clustered-Minds Multiverse Changes Scientific Work and Thinking

Radical Subjectivity and Scientific Progress

The last section already ended with research opportunities—in the domain of decision making. This section will be concentrating on some further, especially exciting research opportunities—without trying to summarize the number of research opportunities already reported in the previous chapters of this book. However, it will start with a more radical question: Looking through the lens of the clustered-minds multiverse, are there actually any possibilities to *gain knowledge* from research? And is there a difference between different scientific domains in this regard?

Chapter 12 has analyzed the consequences of self-fulfilling expectations by experimenters, the consequences of a generalized Rosenthal effect, for the case of the social sciences: The development of scientific knowledge is here limited by *radical subjectivity*. A certain version of the experimenter will be entangled with a certain reality and reports on the outcomes of his experiments in a scientific publication. Sure enough, looking from the perspective of the clustered-minds multiverse, since different versions of the experimenter will be entangled with *different* realities, the experimental results will be different, and publications by those

Table 13.1 Degrees of truth in scientific research

Type of knowledge	Feasible within which type of science
Intersubjective truth within one minds cluster	Social sciences
Intersubjective truth beyond one minds cluster	Physics
Objective truth	Not feasible within any manmade science

different versions will report on different findings to be read in different minds clusters. How much of this thinking is specific to the social sciences and what is the situation with respect to physics?[5] Table 13.1 starts with a simple classification of the type of knowledge that can be generated within the social sciences and physics; this classification will be justified and discussed in the remainder of this subsection.

Let me start with the social sciences. According to Table 13.1, radical subjectivity is not supposed to imply that *no knowledge* can be generated at all in the course of the scientific process, even in the social sciences. As already pointed out in the last chapter, it rather means twice: (a) that the necessity to replicate findings in the social sciences, preferably by various different researchers with various backgrounds, is even more serious an issue than many social scientists would already argue based on a non-quantum worldview and (b) that our knowledge might only be intersubjectively true within a specific minds cluster.

Let me explore this beyond what has been stated in the last chapter and start with (a). Replication is one of the big issues in social psychology within the last few years, partially because of clear cases of fraud that have been detected, partially because of difficulties to replicate well-known results (even if there is no reason at all to assume that the respective primary investigators were fraudulent). People have coined the term 'replication crisis' for this (Schooler 2014; together with a proposal how to address that problem).

An example for a tricky replication situation has already been dealt with in this book. In the last chapter, I have looked at a result on social priming by Bargh et al. (1996) that Doyan et al. (2012) were not able to *fully* replicate; however, based on the clustered-minds multiverse I suggested a different interpretation than the latter authors by arguing that differences between different experiments might be explained by different expectations by the *leading investigators* and might hence be an example for a generalized Rosenthal effect.

As already mentioned, researchers are seriously considering the existence of a general 'replication crisis;' the difficulties to replicate experimental results are not only severe, especially in psychological research, but they are considered threatening for scientific progress by many, perhaps because they cannot fully be explained within our mainstream weltanschauung of a singular, quasi-classical reality. A great example for recent studies on the replicability of psychological results are the several contributions to a special issue of the journal *Social Psychology* (2014). Especially interesting is an article by Klein et al. (2014).

[5]As in other parts of the book, applications beyond the social sciences and physics are not pursued.

According to Klein et al., the results of several studies turn out to be robust, whereas the results of several other studies do not, when put under the scrutiny of replication trials. The question is why, and answering it appears to be hard: "(…) whether the sample was collected in the US or elsewhere, or whether data collection occurred online or in the laboratory, had little systematic effect on the observed results" (Klein et al. 2014, 150). Small versus large samples also did not matter: "(…) most of the variation in effects was due to the effect under investigation," (Klein et al. 2014, 151) i.e., the type of hypothesis tested. Let me mention that this is fully consistent with the idea of generalized experimenter expectancy effects because *those* have *not* systematically been analyzed (anyway impossible with the data available to Klein et al.)—albeit no proof.

My thoughts on (b) can only be speculative since the measurement of a minds cluster would only be possible with a clear empirical (i.e., operational) definition. Many would argue that such an endeavor is hopeless because other realities cannot, anyway, be observed, whereas I have argued at several points in this book that the way out might be indirect measurements, enabled, potentially, via utility interdependencies between versions of individuals in other realities and the version of the individual in the current reality, shared with the researcher. Moreover, the following speculation might not be too far-fetched. Over time, minds clusters might change.[6] Consequently, the membership of certain versions of individuals within a minds cluster is dynamic, minds clusters might frequently be splitting; the mutual agreements within the cluster might also be changing, e.g., the collective agreement on how to perceive reality. Following from this, psychological results might *not* replicate if either the researcher trying to replicate the results of an experiment has different expectations than the researcher that ran the original experiment or if the minds cluster has changed so much that neither enough versions of individuals residing in a certain cluster are 'willing' to replicate the result via a behavior that is consistent with what has been observed in the original study or if the scientific community is not 'willing,' anymore, to accept it.[7] Those effects should on average be larger with *older* findings. An interesting effect to look at would hence be how well old versus recent findings replicate.[8]

Regarding physical experiments, Table 13.1 states the results to be *intersubjectively true* beyond one minds cluster. This is more than what was granted with respect to the knowledge generated within the social sciences but certainly less than what most physicists would hope for. Indeed, Table 13.1 also states that *objective knowledge* is impossible for any manmade science (including physics). How can those classifications be justified? Let me start with the part that physical results

[6]That does not mean that I am now taking any flow of time seriously. Time is one proxy for distance between realities.

[7]This is not to be mistaken with an acceptance by journals etc., albeit this might be part of it.

[8]It is crystal clear that there are many other reasons, too, as to why older results might not replicate as well as more recent ones such as changes in scientific methodology or statistics, changes in the culture of the population the respondents are taken from (this, however, being part of a mind-cluster's setup) etc. Hence, this is no conclusive test of the multiverse-based hypothesis.

should hold across different minds clusters—with its *advantage* over the social sciences. This is surely justified since we suppose that other minds clusters differ from ours solely in their *state of consciousness*—or more precisely with respect to the relative states of the versions of participating individuals—but not with respect to basic 'ingredients' such as the wave function itself, entanglement, the Born rule, special and general relativity etc. So, no matter whether you or a replica of yours carries out some cosmological measurements, you will both find the same results.

At the same time, however, and now turning to the other side of the classification, it is all but clear what the status of those physical measurements or resulting physical theories might be in terms of a '*final truth.*' In a philosophical presentation at a conference on quantum computing, Deutsch (2003) discusses the status of physical knowledge. And although this is a radical example for the debate about the status of physical knowledge, it is a very instructive one.[9] Deutsch (2003) states that the intriguing account that "maybe the universe that we see—or presumably the multiverse—is really a computer program running on a giant computer" (4) is "a fundamentally flawed idea" (4), but his arguments offer no proof for his statement. Let me look more closely at one of his arguments: "(…) [If] what we see as the laws of physics are actually just attributes of some software, then (…) we (…) have no means of understanding the hardware on which that software is running" (4). Since we have no possibility to decide *whether that hardware is actually there or not*, he dismisses this type of reasoning: "(…) [We] have no more reason for postulating that it's there than we have for postulating that there are fairies at the bottom of the garden" (4). From my perspective, Deutsch's comparison of the knowledge status with respect to the giant computer and the knowledge status with respect to fairies is a bit unfair (mostly treating the fairies a bit unfairly …), but this is for the reader to decide. However, following his own reasoning, the absence of a possibility to *prove* that there is *no* such hypothetical, giant computer, does not mean that there is none. I guess that this is simply turning Deutsch's argument on its head, and most probably what we achieve here in this virtual debate is a draw. So even if we would intuitively dismiss the idea of the giant computer or simply dislike it, we are simply not able to decide.

Let me summarize. There are good reasons to do science, not just for the enjoyment of the process, but because science actually generates knowledge. The knowledge we are able to generate within the social sciences is specific to the minds-cluster we are residing in. If we want to move slightly beyond our cluster,[10] numerous replications—by researchers that are as different as possible in their expectations—turn out to be critical. But even physics might not be able to answer questions about the 'final truth,' it might only be able to come up with

---

[9]As already stated in the last chapter, a detailed account of different epistemological positions in philosophy or a thorough analysis of different positions in the philosophy of science is not possible within the scope of this book.

[10]One might argue that this is not possible at all and I would have no stringent argument against that point. However, some realities might differ only smoothly from ours, and we might get a feel for the robustness of our findings via numerous replications.

intersubjective knowledge, knowledge that is important and useful for *mankind*, for how we understand and live our lives, for how we build rockets and cellphones, and for understanding that there is a quantum multiverse 'out there.' Whether we are experiencing our lives within a large computer program or not as well as other possible speculations about the 'final truth' may, however, not be addressed within scientific research.

And let me add one final thought. A great example for the unclear 'knowledge status' of parts of physical theory is the Born (1926) rule. Taking the perspective of the multiverse interpretation, here, I would like to argue that this auxiliary equation, as I called it, somehow accurately predicts a 'random sequence' within one decoherent history, a 'random sequence' that is fully consistent with what will be measured. But I would not be willing to admit that it postulates an *actual* random process to get there; again, the Schrödinger equation is deterministic, and it is fully unclear to me where the randomness might enter the picture. That is the main reason why I was always talking about relative frequencies of measurement outcomes rather than probabilities; but this is perhaps a matter of taste, given the fact that the underlying process generating those relative frequencies is unclear. Therefore, I would like to fully agree with the title (and conclusion) by Landsman (2008), although derived within the framework of a different interpretation of quantum mechanics: "The Conclusion Seems to Be That No Generally Accepted Derivation of the Born Rule Has Been Given to Date, but this Does Not Imply That Such a Derivation Is Impossible in Principle." As already stated, for the several reasons discussed in Chap. 4, the decision-theoretic account by Deutsch, Wallace etc. is not the solution, from my point of view.

A New Type of Psychophysical Experiments

Are there any *novel types of experiments* that should be undertaken, given the considerations within this book? In addition to the exciting, stylized type of experiment on the generalized Rosenthal effect suggested in Chap. 12 and a few more experiments mentioned en passant within previous chapters, I would like to argue that a great way of benefitting from the findings of this book is designing and conducting interdisciplinary experiments that, e.g., link physical research and social science research. *Psychophysics* is a term used within psychological research for much more than a century (starting with Fechner) to describe experiments investigating the covariation of entities in the physical 'sphere' with entities in the psychological 'sphere' such as the *subjective perception* of color or temperature with *physical changes* in color and temperature. Psychophysical parallelism, a concept discussed in Box 5.2 at the end of Chap. 5, has been first suggested within that research paradigm.

But the notion of psychophysics might also be reversed to describe experiments directly investigating the *effect of consciousness within 'subjective physics'* (i.e., top-down decoherence within an open-systems perspective).[11] The starting point might be a certain type of experiment that has been carried out by Radin et al. (2012). Radin et al. (2012) were interested in demonstrating that consciousness has a *direct* influence on processes observed in the subatomic domain, beyond the basics of the measurement problem. It is an example of a study *combining* a psychological and a physical experiment, even bringing in physiological measurements (in one of their experiments). Within six experiments, the authors analyze how directing conscious attention more or less to the system to be measured plays out in terms of the strength of the interference pattern in a double-slit experiment. Indeed, when respondents were told to direct their conscious attention to the double-slit, the strength of the interference pattern was smaller than when they were told to 'relax.' From the perspective of the clustered-minds multiverse, this might be interpreted in a way that directing conscious attention to the double-slit leads to a more and more pronounced *perception* of singular realities and that this transition in perception might be *smooth* on a subatomic level (different from the macro-world situation). Although Radin et al. (2012) try to interpret their results within the framework of the standard, collapse interpretation (i.e., the Copenhagen interpretation, used as an umbrella term), my tentative interpretation shows that those findings might as well be accommodated for within the clustered-minds multiverse. Radin et al.'s (2012) experiments might then become the start of a new paradigm of experimentation linking consciousness and multiverse quantum mechanics.

Other Challenging Topics

Using the above type of psychophysical experiments, we might not only be enabled to understand better empirically the workings of consciousness in the multiverse, but such experiments might also be a starting point for the development of *consciousness theory*. Consciousness theory would help making the *science of consciousness* an independent scientific domain. In turn, novel hypotheses and experiments could be developed. Regarding potential experiments, working from and modifying the above paradigm, one might look at questions such as what happens in *different states of consciousness*, what can be seen in terms of physical reactions, say, at the double slit, if people *sleep*, if they *meditate*, if they *visualize* things etc.?

The decision sciences have thoroughly been analyzed as a research domain that might benefit from utilizing the framework of the clustered-minds-multiverse.

[11]Please recall that within a closed-system perspective (i.e., looking from 'outside' the wave function), there is no effect of conscious measurement on the physical (see Chaps. 2 and 5).

They might, within this framework, also benefit from a cross-fertilization with a science of consciousness. So far, the connection between the decision sciences and the natural sciences has rather worked in favor of a *materialist worldview*. Example are the famous Libet experiments or more recent neuroscience experiments involving individuals' choices (e.g., Soon et al. 2008). I was able to reinterpret the seeming anti-free-will evidence of all those experiments that try to reduce conscious free will to—at most—a vetoing possibility with motor action. Perhaps, the same type of rethinking is also a starting point for a fresh look into the connection between the neurosciences, consciousness and decision making more generally.

In fact, there is hardly any domain of science that would not require some rethinking when a multiverse perspective is applied. In this book, I have restricted the considerations to physics and the social sciences. But what would, e.g., happen to evolution theory when applying a multiverse perspective? What is even *survival* when there is a number of versions of each living entity? I am not indulging in 'attention-getters' like the 'quantum suicide,' here, I am rather talking about perspectives in tune with the relationship that Mensky (2010) sees between the *anthropic principle* and the multiverse. Translated into my theoretical framework, individuals might be able to allocate more consciousness to realities where they will survive. Turning this into a new version of evolution theory would require to define what types of animals are able to make vectorial choices: apes, cats, dolphins, rats (or even all animals)? And how could one actually tell? Rats taking measurements at the double slit? This is just a humorous, but perhaps revealing example for the challenging type of discussions lying ahead when pursuing a consequent implementation of the perspective offered by the clustered-minds multiverse within future research in several fields of science. This is not meant to be frustrating. It rather shows what is asked from researchers engaging in this exciting project, independent of the field of research they are working in: openness, tolerance and no dogmatism.

References

Abhayawansa, Kapila. 2013. "Buddhism and Moral Responsibility: Response to Shyamon." *Colombo Telegraph*. August 1, 2013. Accessed October 23, 2015. https://www.colombotelegraph.com/index.php/buddhism-and-moral-responsibility-response-to-shyamon/.

Abric, Jean-Claude, and James P. Kahan. 1972. "The Effects of Representations and Behavior in Experimental Games." *European Journal of Social Psychology* 2: 129–144.

Adlam, Emily. 2014. "The problem of confirmation in the Everett interpretation." *Studies in History and Philosophy of Science Part B: Studies in History and Philosophy of Modern Physics* 47: 21–32.

Aharanov, Yakir, Peter G. Bergmann, and Joel L. Lebowitz. 1964. "Time Symmetry in the Quantum Process of Measurement." *Physical Review* 134: B1410-B1416.

Albert, David, and Barry Loewer. 1988. "Interpreting the Many Worlds Interpretation." *Synthese* 77: 195–213.

Alexandrova, Anna. 2005. "Subjective Well-Being and Kahneman's 'Objective Happiness.'" *Journal of Happiness Studies* 6: 301–324.

Allais, Maurice. 1953. "Le Comportement de l'Homme Rationnel Devant Le Risque: Critique des Postulats et Axiomes de L'Ecole Americaine." *Econometrica* 21: 503–546.

Ammons, Robert B., and Carol H. Ammons. 1957. "ESP and PK: A Way Out?" *North Dacota Quarterly* 25: 119–121.

Andersen, Richard A., and Stephen M. Kosslyn, eds. 1992. *Frontiers in Cognitive Neuroscience*. Cambridge, MA: MIT press.

Anderson, Edward. 2012. "Problem of Time in Quantum Gravity." *Annalen der Physik* 524: 757–786.

Aquinas, Saint Thomas. (1265–1273) 2006. "Summa Theologiae: God's Will and Providence." In Vol. 5 of *The Dominican Council*, edited by Thomas Gilby, 19–26. Cambridge: Cambridge University Press.

Aristotle. (~ 350 B.C.) 1999. *Aristotle Physics, Book VIII*, edited by Daniel W. Graham. Oxford: Clarendon Press.

Aristotle. (~ 350 B.C.) 1985. *The Nicomachean Ethics*, translated by Terence Irwin. Indianapolis: Hackett Publishing Co.

Auletta, Gennaro. 2001. *Foundations and Interpretation of Quantum Mechanics*. Singapore: World Scientific Publishing Co.

Bacciagaluppi, Guido. 2001. "Remarks on Space-Time and Locality in Everett's Interpretation." Talk delivered at the NATO Advanced Research Workshop on Modality, Probability, and Bell's Theorems, Cracow, August 19–23. Accessed April 17, 2018. https://www.philsci-archive.pitt.edu/504/.

© Springer Nature Switzerland AG 2018

C. D. Schade, *Free Will and Consciousness in the Multiverse*,
https://doi.org/10.1007/978-3-030-03583-9

Barbour, Julian. 1999. *The End of Time: The Next Revolution in our Understanding of the Universe*. UK: Weidenfeld & Nicholson.

Bargh, John A., Mark Chen, and Lara Burrows. 1996. "Automaticity of Social Behavior: Direct Effects of Trait Construct and Stereotype-Activation on Action." *Journal of Personality and Social Psychology* 71: 230–244.

Barrett, Jeffrey A. 1999. *The Quantum Mechanics of Minds and Worlds*. Oxford: Oxford University Press.

Beck, Friedrich, and John C. Eccles. 1992. "Quantum Aspects of Brain Activity and the Role of Consciousness." *Proceedings of the National Academy of Science USA* 89: 11357–11361.

Becker, Gary S. 1974. "Crime and Punishment: An Economic Approach." In *Essays in the Economics of Crime and Punishment*, edited by Gary S. Becker and William M. Landes. Cambridge, MA: National Bureau of Economic Research.

Bell, John. 1964. "On the Einstein-Podolsky-Rosen Paradox." *Physics* 1: 195–200.

Beller, Mara. 1999. *Quantum Dialogue: The Making of a Revolution*. Chicago: The University of Chicago Press.

Bentham, Jeremy. (1789) 1996. *An Introduction to the Principles of Morals and Legislation*. London: T. Payne and Son. Reprint, edited by James H. Burns and Herbert L. A. Hart. Oxford: Clarendon Press.

Bernoulli, Daniel. (1738) 1954. "Exposition of a New Theory on the Measurement of Risk," translated by Dr. Louise Sommer. *Econometrica* 22: 22–36.

Bohm, David. 1952. "A Suggested Interpretation of the Quantum Theory in Terms of 'Hidden Variables' I." *Physical Review* 85: 166–179.

Boltzmann, Ludwig. 1895. "On Certain Questions of the Theory of Gases." *Nature* 51: 413–415.

Born, Max. 1926. "Zur Quantenmechanik der Stoßvorgänge." *Zeitschrift für Physik* 37: 863–867.

Boscá Díaz-Pintado, María C. 2007. "Updating the Wave-Particle Duality." Paper presented at the 15th UK and European Meeting on the Foundations of Physics, Leeds, March 29–31.

Boscá, María C. 2009: "Representaciones en microfísica: Sobre la mecánica cuántica." In *Varianciones Representionales: Entre lo literal y lo translaticio*, edited by Martínez Freire, Pascual. *Contrastes* 14: 177–198.

Broad, Charlie D. 1934. *Determinism, Indeterminism, and Libertarianism*. London: Cambridge University Press. (Reprinted in Charlie D. Broad. 1952. *Ethics and the History of Philosophy*. London: Routledge & Kegan Paul, 195-217.)

Burmeister, Katrin, and Christian D. Schade. 2007. "Are entrepreneurs' decisions more biased? An experimental investigation of the susceptibility to status quo bias." *Journal of Business Venturing* 22: 340–362.

Busemeyer, Jerome R., Zheng Wang, and Ariane Lambert-Mogiliansky. 2009. "Empirical Comparison of Markov and Quantum Models of Decision Making." *Journal of Mathematical Psychology* 53: 423–433.

Caillaud, Bernard, and Bruno Jullien. 2003. "Chicken and Egg: Competition among Intermediation Service Providers." *Rand Journal of Economics* 34: 309–328.

Camerer, Colin, and Dan Lovallo. 1999. "Overconfidence and Excess Entry: An Experimental Approach." *American Economic Review* 89: 306–318.

Cannon, Walter B. 1942. "Voodoo Death." *American Anthropologist* 44: 169–181.

Chalmers, David J. 1995. "Facing Up to the Problem of Consciousness." *Journal of Consciousness Studies* 2: 200–219.

Chalmers, David J. 1996. *The Conscious Mind in Search of a Fundamental Theory*. Oxford: Oxford University Press.

Chalmers, David J. 2010. *The Character of Consciousness*. Oxford: Oxford University Press.

Chu, James A. 1991. "The Repetition Compulsion Revisited: Reliving Dissociated Trauma." *Psychotherapy* 28: 327–332.

Cohen, Jacob. 1977. *Statistical Power Analysis for the Behavioral Sciences*, rev. ed. New York: Academic Press.

Collins, Robert. 2011. "The Energy of the Soul." In *The Soul Hypothesis: Investigations into the Existence of the Soul*, edited by Mark C. Baker and Steward Goetz, 123–137. New York, London: Continuum International Publishing Group.
Dasgupta, Surendranath. 1962. *Indian Idealism*. London: Cambridge University Press.
Davidson, Donald. (1970) 1980. "Mental Events," reprinted in *Essays on Actions and Events*, edited by Donald Davidson, 207–225. Oxford: Clarendon Press.
Dennett, Daniel C. 1978. *Brainstorms: Philosophical Essays on Mind and Psychology*. Boston: MIT Press.
Dennett, Daniel C. 1991. *Consciousness Explained*. Boston: Little Brown.
Dennett, Daniel C. 2003a. "The Illusion of Consciousness." Ted talk 2003. Accessed March 12, 2015. http://www.ted.com/talks/dan_dennett_on_our_consciousness.
Dennett, Daniel C. 2003b. *Freedom Evolves*. New York: Viking Press.
Deutsch, David. 1985. "Quantum Theory as a Universal Physical Theory." *International Journal of Theoretical Physics* 24: 1–41.
Deutsch, David. 1991. "Quantum Mechanics near Closed Timelike Curves." *Physical Review* 44: 3197–3217.
Deutsch, David. 1997. *The Fabric of Reality: Towards a Theory of Everything*. Middlesex: Penguin Books Ltd.
Deutsch, David. 1999. "Quantum Theory of Probability and Decisions." *Proceedings of the Royal Society of London* 455: 3129–3137. https://doi.org/10.1098/rspa.1999.0443 .
Deutsch, David. 2003. "Physics, Philosophy and Quantum Technology." In *Proceedings of the 6th International Conference on Quantum Communication, Measurement and Computing*, edited by Jeffrey Shapiro and Osamu Hirota. 2002: Massachusetts Institute of Technology. Princeton, NJ: Rinton Press. Accessed online December 17, 2017. http://gretl.ecn.wfu.edu/~cottrell/OPE/archive/0305/att-0257/02-deutsch.pdf. Citations refer to the online version.
Deutsch, David. 2011. "Vindication of Quantum Locality." *Proceedings of the Royal Society A* 468: 531–544.
Deutsch, David. 2012a. "Apart from Universes." In *Many Worlds? Everett, Quantum Theory, & Reality*, edited by Simon Saunders, Jonathan Barrett, Adrian Kent, and David Wallace, 542–552. Oxford: Oxford University Press.
Deutsch, David. 2012b. *The Beginning of Infinity: Explanations that Transform the World*. Middlesex: Penguin Books Ltd.
Deutsch, David. 2016. "The Logic of Experimental Tests, Particularly of Everettian Quantum Theory." *Studies in History and Philosophy of Modern Physics* 55: 24–33.
Deutsch, David, and Patrick Hayden. 2000. "Information Flow in Entangled Quantum Systems." *Proceedings of the Royal Society A* 456: 1759–1774.
Deutsch, David, and Michael Lockwood. 1994. "The Quantum Physics of Time Travel." *Scientific American* 270: 68–74.
Dew, Nicholas, Stuart Read, Saras D. Sarasvathy, and Robert Wiltbank. 2009. "Effectual Versus Predictive Logics in Entrepreneurial Decision-Making: Differences Between Experts and Novices." *Journal of Business Venturing* 24: 287–309.
DeWitt, Bryce S. 1967. "Quantum Theory and Gravity. I. The Canonical Theory." *Physical Review* 160: 1113–1148.
DeWitt, Bryce S. 1970. "Quantum Mechanics and Reality: Could the Solution to the Dilemma of Indeterminism Be a Universe in Which All Possible Outcomes of an Experiment Actually Occur?" *Physics Today* 23: 30–40.
DeWitt, Bryce S. 1971. "The Many-Universes Interpretation of Quantum Mechanics." In *Foundations of Quantum Mechanics*, edited by Bernard D'Espagnat. New York: Academic Press.
Diener, Ed, Robert A. Emmons, Randy J. Larsen, and Sharon Griffin. 1985. "The Satisfaction with Life Scale." *Journal of Personality Assessment* 49: 71–75.

Diener, Ed, and Richard E. Lucas. 1999. "Personality and Subjective Well-Being." In *Well-Being: The Foundations of Hedonic Psychology*, edited by Daniel Kahneman, Ed Diener, and Norbert Schwarz. New York: Russel Sage Foundation.

Dixit, Avinash K. 1992. "Investment and Hysteresis." *Journal of Economic Perspectives* 6: 107–132.

Dixit, Avinash K., and Robert S. Pindyck. 1994. *Investment Under Uncertainty*. Princeton University Press.

Dowker, Fay, and Adrian Kent. 1996. "On the Consistent Histories Approach to Quantum Mechanics." *Journal of Statistical Physics* 82: 1557–1646.

Doyen, Stéphane, Olivier Klein, Cora-Lise Pichon, and Axel Cleeremans. 2012. "Behavioral Priming: It's All in the Mind, but Whose Mind?" *PlosOne* 7: 1–7. Accessed December 22, 2015. https://www.plosone.org , e29081.

Eccles, John C. 1985. "Mental Summation: The Timing of Voluntary Intentions by Cortical Activity." *Behavioral and Brain Sciences* 8: 542–543.

Eckel, Catherine C., and Philip J. Grossman. 1996. "Altruism in Anonymous Dictator Games." *Games and Economic Behavior* 16: 181–191.

Edgeworth, Francis Y. (1881) 1967. *Mathematical Psychics: An Essay on the Application of Mathematics to the Moral Sciences*. Reprint. New York: M. Kelly.

Efron, Bradley. 1979. "Bootstrap Methods: Another Look at Jacknife." *The Annals of Statistics* 7: 1–26.

Einstein, Albert, Boris Podolsky, and Nathan Rosen. 1935. "Can Quantum-Mechanical Description of Physical Reality Be Considered Complete?" *Physical Review* 47: 777–780.

Eisenstein, Charles. 2011. *Sacred Economics: Money, Gift and Society in the Age of Transition*. Berkeley: Evolver Editions.

Elitzur, Avshalom C., Eliahu Cohen, and Tomer Shushi. 2016. "The Too-Late-Choice Experiment: Bell's Proof Within a Setting Where the Nonlocal Effect's Target is an Earlier Event." *International Journal of Quantum Foundations* 2: 32–46. arXiv:1512.08275v2.

Ellsberg, Daniel. 1961. "Risk, Ambiguity and the Savage Axioms." *Quarterly Journal of Economics* 75: 643–669.

Engel, Christoph. 2010. "Dictator Games: A Meta Study." Preprints of the Max Planck Institute for Research on Collective Goods. Bonn 2010/07. Accessed December 12, 2015. http://www.coll.mpg.de/pdf_dat/2010_07online.pdf.

Eshleman, Andrew. 2014. "Moral Responsibility." In *Stanford Encyclopedia of Philosophy*. Accessed October 22, 2015. http://plato.stanford.edu/entries/moral-responsibility/index.html#note-1.

Everett, Hugh, III. 1957. "'Relative State' Formulation of Quantum Mechanics." *Reviews of Modern Physics* 29: 454–462.

Everett, Hugh, III. 1973. "The Theory of the Universal Wave Function." Ph.D. thesis, Princeton University. In *The Many-Worlds Interpretation of Quantum Mechanics*, edited by Bryce S. DeWitt and Neill Graham, 3–140. Princeton, NJ: Princeton University Press.

Farmer, Roger E. A., and Michael Woodford. 1997. "Self-Fulfilling Prophecies and the Business Cycle." *Macroeconomic Dynamics* 1: 740–769.

Faschingbauer, Michael. 2013. *Wie erfolgreiche Unternehmer denken, entscheiden und handeln*. 2nd, rev. ed. containing additional material. Stuttgart: Schäffer-Poeschel.

Fechner, Gustav T. 1860. *Elemente der Psychophysik*. 2 vols. Leipzig: Breitkopf & Härtel.

Feynman, Richard. 1967. *The Character of Physical Law*. London: The British Broadcasting Corporation. Reprint. Cambridge and London: The MIT Press.

Fishbein, Martin, and Icek Ajzen. 1975. *Belief, Attitude, Intention, and Behavior: An Introduction to Theory and Research*. Reading, MA: Addison-Wesley.

Fischer, John M., and Mark Ravizza. 1992. "When the Will is Free." *Philosophical Perspectives* 6: 423–451.

Frankfurt, Harry. 1969. "Alternate Possibilities and Moral Responsibility." *Journal of Philosophy* 66: 829–839.

Frauchinger, Daniela, and Renato Renner. 2016. "Single-World Interpretations of Quantum Theory Cannot Be Self-Consistent." Working Paper. arXiv:1604.07422v1.

Freud, Anna. 1937. *The Ego and the Mechanisms of Defense.* London: Hogarth Press and Institute of Psychoanalysis.

Freud, Sigmund. 1917. "Eine Schwierigkeit der Psychoanalyse." *Imago. Zeitschrift für Anwendung der Psychoanalyse auf die Geisteswissenschaften* 5: 1–7.

Freud, Sigmund. 1955. "Beyond the Pleasure Principle." In vol. XVIII of *The Standard Edition of the Complete Psychological Works of Sigmund Freud.* Translated from the German under the general editorship of James Strachey in collaboration with Anna Freud (and assisted by others). London: The Hogarth Press and the Institute of Psychoanalysis. (Originally published as Freud, Sigmund. 1920. *Jenseits des Lustprinzips.* Leipzig et al.: Internationaler Psychoanalytischer Verlag.)

Freud, Sigmund. 2010. *The Interpretation of Dreams.* Strachey, James. New York: Basic Books. (Originally published as Freud, Sigmund. 1900. *Die Traumdeutung*. Leipzig, Wien: Franz Deuticke.)

Friedman, Lawrence. 2019. *Freud's Papers on Technique and Contemporary Clinical Practice.* London and New York: Routledge.

Fuchs, Christopher A. 2010. "QBism: The Perimeter of Quantum Baysianism." Accessed April 21, 2014. http://arxiv.org/abs/1003.5209.

Fumerton, Richard A. 2006. "Solipsism." In vol. 9 of *Encyclopedia of Philosophy*, 2nd ed., edited by Donald M. Borchert, 115–122. Detroit: Macmillan Reference USA.

Gale, Richard. 1966. "McTaggart's Analysis of Time." *American Philosophical Quarterly* 3: 145–152.

Galvan, Bruno. 2010. "On the Preferred-Basis Problem and It's Possible Solutions." arXiv:1008.3708v1[quant-ph].

Gazzaniga, Michael, Richard B. Ivry, and George R. Mangun. 2013. *Cognitive Neuroscience: The Biology of the Mind,* 4th ed. New York: W. W. Norton.

Gigerenzer, Gerd, and Peter M. Todd. 1999. "Ecological Rationality: The Normative Study of Heuristics." In *Ecological Rationality. Intelligence in the World*, edited by Gerd Gigerenzer, Peter M. Todd, and the ABC Research Group, 487–497. New York: Oxford University Press.

Gigerenzer, Gerd, Peter M. Todd, and the ABC Research Group, eds. 1999. *Ecological Rationality. Intelligence in the World.* New York: Oxford University Press.

Ginet, Carl. 1997. "Freedom, Responsibility, and Agency." *The Journal of Ethics* 1: 85–98.

Glasersfeld, Ernst von. 1984. "An Introduction to Radical Constructivism." In *The Invented Reality: How Do We Know What We Believe We Know? (Contributions to Constructivism)*, edited by Paul Watzlawick, 17–40. New York: Norton.

Goetz, Steward. 2011. "Making Things Happen: Souls in Action." In *The Soul Hypothesis: Investigations into the Existence of the Soul*, edited by Mark C. Baker and Steward Goetz, 99–122. New York, London: Continuum International Publishing Group.

Goldstein, Jeffrey. 1999. "The Attraction of Violent Entertainment." *Media Psychology* 1: 271–282.

Goldstein, Jeffrey, ed. 1998. *Why We Watch. The Attractions of Violent Entertainment.* New York: Oxford University Press.

Goswami, Amit. 2015. *Quantum Economics. Unleashing the Power of an Economics of Consciousness.* Faber, VA: Rainbow Ridge Books.

Greaves, Hilary. 2004. "Understanding Deutsch's Probability in a Deterministic Multiverse." *Studies in History and Philosophy of Modern Physics* 35: 423–456.

Griffin, Donald R., and Gayle B. Speck. 2004. "New Evidence of Animal Consciousness." *Animal Cognition* 7: 5–18.

Grosholz, Emily. 2011. "Reference and Analysis: The Representation of Time in Galileo, Newton, and Leibniz. 2010 Arthur O. Lovejoy Lecture." *Journal of the History of Ideas* 72: 333–350.

Gruber, Ronald, and Richard A. Block. 2012. "Experimental Evidence That the Flow of Time Is a Perceptual Illusion." Paper presented at the "Toward a Science of Consciousness" conference, Tucson, AZ, April 9-14, 2012.

Güth, Werner, Rolf Schmittberger, and Bernd Schwarze. 1982. "An Experimental Analysis of Ultimatum Bargaining." *Journal of Economic Behavior and Organization* 3: 367–388.

Guthrie, Edwin R. 1938. *The Psychology of Human Conflict*. New York: Harper and Row.

Halvorson, Hans. 2011. "The Measure of All Things: Quantum Mechanics and the Soul." In *The Soul Hypothesis: Investigations into the Existence of the Soul*, edited by Mark C. Baker and Steward Goetz, 138–167. New York, London: Continuum International Publishing Group.

Hameroff, Stuart R. 2012. "How Quantum Biology Can Rescue Conscious Free Will." *Frontiers in Integrative Neuroscience* 6, article 93:1–17. https://doi.org/10.3389/fnint.2012.00093.

Hameroff, Stuart R., and Roger Penrose. 1995. "Orchestrated Reduction of Quantum Coherence in Brain Microtubules: A Model for Consciousness." *Neural Network World* 5: 793–804.

Hansen, Chad. 1972. "Freedom and Moral Responsibility in Confucian Ethics." *Philosophy East and West* 22: 169–186.

Hare, Richard M. 1984. "Supervenience." Inaugural address to the Aristotelian Society. *Aristotelian Society Supplementary* 58: 1–16.

Harrison, Glenn W., and E. Elisabet Rutström. 2015. "Risk Aversion in the Laboratory." In *Risk Aversion in Experiments* (*Research in Experimental Economics*, vol. 12), edited by James C. Cox and Glenn W. Harrison, 41–196. Bingley: Emerald Group Publishing Limited. http://dx.doi.org/10.1016/S0193-2306(08)00003-3 .

Harsanyi, John C., and Reinhard Selten. 2003. *A General Theory of Equilibrium Selection in Games*. Cambridge, MA: MIT Press.

Haven, Emmanuel, and Andrei Khrennikov. 2013. *Quantum Social Science*. Cambridge: Cambridge University Press.

Hayes, Nicky. 1994. *Foundations of Psychology*. New York: Routledge.

Heath, Chip, and Amos Tversky. 1991. "Preference and Belief: Ambiguity and Competence in Choice under Uncertainty." *Journal of Risk and Uncertainty* 4: 5–28.

Heisenberg, Werner. 1927. "Über den anschaulichen Inhalt der quantentheoretischen Kinematik und Mechanik." *Zeitschrift für Physik* 43: 172–198. (English translation in Wheeler and Zurek: Heisenberg, Werner. 1927. "The Physical Content of Quantum Kinematics and Mechanics," 62–84.) Page references refer to Wheeler and Zurek.

Herdova, Marcela. 2016. "What You Don't Know *Can* Hurt You: Situationism, Conscious Awareness, and Control." *Journal of Cognition and Neuroethics* 4 (1): 45–71.

Herzog, Thomas J., Paul G. Kwiat, Harald Weinfurter, and Anton Zeilinger. 1995. "Complementarity and the Quantum Eraser." *Physical Review Letters* 75: 3034–3037.

Holt, Charles A., and Susan K. Laury. 2002. "Risk Aversion and Incentive Effects." *American Economic Review* 92: 1644–1655.

Hume, David. (1740) 1967. *A Treatise of Human Nature. Section VIII.: Of Liberty and Necessity*. Oxford: Oxford University Press.

Jönsson, Claus. 1961. "Elektroneninterferenzen an mehreren künstlich hergestellten Feinspalten." *Zeitschrift für Physik* 161: 454–474.

Joos, Erich, H. Dieter Zeh, Claus Kiefer, Domenico Giulini, Joachim Kupsch, and Ion-Olimpiu Stamatescu. 2003. *Decoherence and the Appearance of a Classical World in Quantum Theory*. Berlin-Heidelberg: Springer-Verlag.

Jussim, Lee, and Jacquelynne Eccles. 1995. "Naturally Occurring Interpersonal Expectancies." In *Social Development. Review of Personality and Social Psychology*, edited by Nancy Eisenberg, 15: 74–108. Thousand Oaks, CA: Sage Publications.

Kahn, Barbara E., Manohar U. Kalwani, and Donald G. Morrison. 1986. "Measuring Variety-Seeking and Reinforcement Behaviors Using Panel Data." *Journal of Marketing Research* 23: 89–100.

Kahn, Barbara E., and Alice M. Isen. 1993. "The Influence of Positive Affect on Variety Seeking among Safe, Enjoyable Products." *Journal of Consumer Research* 20: 257–270.

Kahneman, Daniel. 1988. "Experimental Economics: A Psychological Perspective." In *Bounded Rational Behavior in Experimental Games and Markets*, edited by Reinhard Tietz, Wulf Albers, and Reinhard Selten, 11–18. Berlin: Springer-Verlag.

Kahneman, Daniel. 2000. "Experienced Utility and Objective Happiness: A Moment-Based Approach." In Chapter 37 of *Choices, Values and Frames*, edited by Daniel Kahneman and Amos Tversky, 673-692. New York: Cambridge University Press and the Russell Sage Foundation.

Kahneman, Daniel. 2011. *Thinking, Fast and Slow*. New York: Farrar, Straus and Giroux.

Kahneman, Daniel, and Robert Sugden. 2005. "Experienced Utility as a Standard of Policy Evaluation." *Environmental & Resource Economics* 32: 161–181.

Kahneman, Daniel, and Amos Tversky. 1979. "Prospect Theory: An Analysis of Decision under Risk." *Econometrica* 47: 263–292.

Kahneman, Daniel, Jack L. Knetsch, and Richard H. Thaler. 1986. "Fairness and the Assumptions of Economics." *Journal of Business* 59: S285-S300.

Kahneman, Daniel, Jack L. Knetsch, and Richard H. Thaler. 1991. "Anomalies: The Endowment Effect, Loss Aversion, and Status Quo Bias." *Journal of Economic Perspectives* 5: 193–206.

Kahneman, Daniel, Peter P. Wakker, and Rakesh Sarin. 1997. "Back to Bentham? Explorations of Experienced Utility." *The Quarterly Journal of Economics* 112: 375–405.

Kane, Robert H. 1985. *Free Will and Values*. Albany, NY: State University of New York Press.

Kane, Robert H. 2003. "Free Will: New Directions for an Ancient Problem." In *Free Will*, edited by Robert H. Kane, 222–248. Oxford: Blackwell.

Kane, Robert H. 2005. *A Contemporary Introduction to Free Will*. New York: Oxford University Press.

Kane, Robert H. 2015. "Free Will: An Achievement Over Indeterminacy." Accessed October 15, 2015. http://www.slate.com/bigideas/are-we-free/essays-and-opinions/robert-kane-opinion.

Kant, Immanuel. (1781) 1996. *Critique of Pure Reason*. Unified ed. Indianapolis, IN: Hackett Publishing Company.

Kauder, Emil. 1953. "Genesis of the Marginal Utility Theory: From Aristotle to the End of the Eighteenth Century." *The Economic Journal* 63: 638–650.

Kauffman, Stuart A. 2010. *Reinventing the Sacred. A New View of Science, Reason, and Religion*. New York: Basic Books.

Kiesler, Sara, Lee Sproull, and Keith Waters. 1996. "A Prisoner's Dilemma Experiment on Cooperation with People and Human-Like Computers." *Journal of Personality and Social Psychology* 70: 47–65.

Kim, Jaegwon. (1984) 1993. *Concepts of Supervenience*. Reprint. Cambridge: Cambridge University Press.

Kitron, David G. 2003. "Repetition Compulsion and Self-Psychology: Towards a Reconciliation." *The International Journal of Psychoanalysis* 84: 427–441.

Klein, Richard A., Kate A. Ratliff, Michelangelo Vianello, Reginald B. Adams, Jr., Štěpán Bahník, Michael Jason Bernstein, Konrad Bocian, et al. 2014. "Investigating Variation in Replicability: A "Many Labs" Replication Project." *Social Psychology* 45: 142–152.

Kleindorfer, Paul R., Howard C. Kunreuther, and Paul J. H. Schoemaker. 1993. *Decision Sciences: An Integrative Perspective*. Cambridge et al.: Cambridge University Press.

Kripke, Saul. (1972) 1980. *Naming and Necessity*. Boston: Harvard University Press.

Kubie, Lawrence S. 1939. "A Critical Analysis of the Concept of a Repetition Compulsion." *The International Journal of Psychoanalysis* 20: 390–402.

Kuczynski, John-M. 2004. "A Quasi-Materialist, Quasi-Dualist Solution to the Mind-Body Problem." *Kriterion: Journal of Philosophy* 45: 81–135.

Kuczynski, John-M. 2015. *The Mind-Body Problem: Philosophy Shorts Volume 21*. Kindle edition.

Kühn, Simone, and Marcel Brass. 2009. "Retrospective Construction of the Judgement of Free Choice." *Consciousness and Cognition* 18: 12–21.

Kuhn, Thomas S. (1962) 1996. *The Structure of Scientific Revolutions*. 3rd ed. Chicago: The University of Chicago Press.

Kunreuther, Howard, Robert Meyer, Richard Zeckhauser, Paul Slovic, Barry Schwartz, Christian D. Schade, Mary-Frances Luce, et al. 2002. "High Stakes Decision Making: Normative, Descriptive and Prescriptive Considerations." *Marketing Letters* 13: 259–268.

Lancaster, Kelvin J. 1966. "A New Approach to Consumer Theory." *Journal of Political Economy* 74: 132–157.

Landsman, Nicholas P. 2008. "The Conclusion Seems to Be That No Generally Accepted Derivation of the Born Rule Has Been Given to Date, but this Does Not Imply That Such a Derivation Is Impossible in Principle." In *Compendium of Quantum Physics*, edited by Friedel Weinert, Klaus Hentschel, Daniel Greenberger, and Brigitte Falkenburg. Berlin: Springer-Verlag.

Lefton, Lester A. 1994. *Psychology*. Needham Heights, MA: Allyn & Bacon.

Leibniz, Gottfried W. (1714) 1898. *The Monadology,* translated by Robert Latta. Accessed January 12, 2018 at http://home.datacomm.ch/kerguelen/monadology/.

Levy, Michael S. 2000. "A Conceptualization of the Repetition Compulsion." *Psychiatry* 63: 45–63.

Lewis, Clarence I. (1929) 1956. *Mind and the World Order. Outline of a Theory of Knowledge*. 1st ed. New York: Dover Publications.

Lewis, David. 1994. "Reduction of Mind." In *A Companion to the Philosophy of Mind*, edited by Samuel Guttenplan, 412–431. Oxford: Blackwell.

Li, Shanjun, and Yiyi Zhou. 2015. "Dynamics of Technology Adoption and Critical Mass: The Case of the U.S. Electric Vehicle Market (September 30, 2015)." NET Institute Working Paper 15-10. Accessed December 19, 2015. http://ssrn.com/abstract=2672065 .

Libet, Benjamin. 1985. "Unconscious Cerebral Initiative and the Role of Conscious Will in Voluntary Action." *The Behavioral and Brain Sciences* 8: 529–539.

Libet, Benjamin. 1999. "Do We Have Free Will?" *Journal of Consciousness Studies* 6: 47–57.

Libet, Benjamin, Elwood W. Wright, Jr., and Curtis A. Gleason. 1982. "Readinesspotentials Preceding Unrestricted 'Spontaneous' vs. Pre-Planned Voluntary Acts." *Electroencephalography and Clinical Neurophysiology* 54: 322–335.

Libet, Benjamin, Curtis A. Gleason, Elwood W. Wright, Jr., and Dennis K. Pearl. 1983. "Time of Conscious Intention to Act in Relation to Onset of Cerebral Activities (Readiness-Potential): The Unconscious Initiation of a Freely Voluntary Act." *Brain* 106: 623–642.

Lillie, Ralph S. 1927. "Physical Indeterminism and Vital Action." *Science* 66: 139–144.

Lockwood, Michael. 1991. *Mind, Brain and the Quantum: The Compound "I."* Oxford: Basil Blackwell. Paperback edition.

Lockwood, Michael. 1996. "'Many Minds' Interpretations of Quantum Mechanics." *British Journal of the Philosophy of Science* 47: 159–188.

Loomes, Graham, and Robert Sugden. 1982. "Regret theory: An Alternative Theory of Rational Choice Under Uncertainty." *Economic Journal* 92: 805–824.

Lupien, Sonia J., Françoise S. Maheu, Mai Thanh Tu, Alexandra J. Fiocco, and Tania E. Schramek. 2007. "The Effects of Stress and Stress Hormones on Human Cognition: Implications for the Field of Brain and Cognition." *Brain & Cognition* 65: 209–237.

Mach, Ernst. 1883. *Die Mechanik in ihrer Entwicklung. Historisch-kritisch dargestellt.* Leipzig: J. A. Barth. (English translation: Mach, Ernst. 1960. The Science of Mechanics: A Critical and Historical Account of its Development. LaSalle, IL: Open Court.)

Machina, Mark J. 1982. "'Expected Utility' Analysis Without the Independence Axiom." *Econometrica* 50: 277–323.

Mas-Colell, Andreu, and Paulo K. Monteiro. 1996. "Self-Fulfilling Equilibria: An Existence Theorem for a General State Space." *Journal of Mathematical Economics* 26: 51–62.

McCabe, Kevin, Daniel Houser, Lee Ryan, Vernon Smith, and Theodore Trouard. 2001. "A Functional Imaging Study of Cooperation in Two-Person Reciprocal Exchange." *Proceedings of the National Academy of Sciences* 98: 11832–11835.

McKenna, Michael, and D. Justin Coates. 2015. "Compatibilism." In *Stanford Encyclopedia of Philosophy*. Accessed October 12, 2015. http://plato.stanford.edu/entries/compatibilism/.

McLaughlin, Brian P. 1984. "Perception, Causation, and Supervenience." *Midwest Studies in Philosophy* 9: 569–592.
McLaughlin, Brian P. 1995. "Varieties of Supervenience." In *Supervenience: New Essays*, edited by Elias Savellos and Ümit Yalcin, 16–59. Cambridge: Cambridge University Press.
McTaggart, John M. E. 1908. "The Unreality of Time." *Mind* 17: 457–473.
Mensky, Michael. B. 2000a. "Quantum Mechanics: New Experiments, New Applications, and New Formulations of Old Questions." *Physics – Uspekhi* 43: 585–600.
Mensky, Michael. B. 2000b. *Quantum Measurement and Decoherence: Models and Phenomenology*. Dordrecht (NL): Springer Science and Business Media.
Mensky, Michael. B. 2001. "Quantum Measurement: Decoherence and Consciousness." Letter to the editor. *Physics – Uspekhi* 44: 438–442.
Mensky, Michael. B. 2005. "Concept of Consciousness in the Context of Quantum Mechanics." *Physics – Uspekhi* 48: 389–409.
Mensky, Michael. B. 2007a. "Quantum Measurements, the Phenomenon of Life, and Time Arrow: The Great Problems of Physics (in Ginzburg's Terminology) and Their Interrelation." *Physics – Uspekhi* 50: 397–407.
Mensky, Michael. B. 2007b. "Postcorrection and Mathematical Model of Life in Extended Everett's Concept." *NeuroQuantology* 5: 363–376.
Mensky, Michael. B. 2010. *Consciousness and Quantum Mechanics: Life in Parallel Worlds*. Singapore: World Scientific Publishing Co.
Mensky, Michael. B. 2011. "Mathematical Models of Subjective Preferences in Quantum Concept of Consciousness." *NeuroQuantology* 9: 614-620.
Mensky, Michael. B. 2013. "Everett Interpretation and Quantum Concept of Consciousness." *NeuroQuantology* 11: 85–96.
Menzel, Ralf, Dirk Puhlmann, Axel Heuer, and Wolfgang P. Schleich. 2012. "Wave-Particle Dualism and Complementarity Unraveled by a Different Mode." *Proceedings of the National Academy of Sciences* 109: 9314–9319. https://doi.org/10.1073/pnas.1201271109 .
Merton, Robert K. 1948. "The Self-Fulfilling Prophecy." *The Antioch Review* 8: 193–210.
Minkowski, Hermann. 1952. "Space and Time." In *The Principle of Relativity: A Collection of Original Memoirs on the Special and General Theory of Relativity*, edited by Hendrik A. Lorentz, Albert Einstein, Hermann Minkowski, and Hermann Weyl, 75–91. New York: Dover. (First presented at the 80th meeting of natural scientists, Cologne, Germany, Sept. 21, 1908.)
Mittelstaedt, Peter, A. Prieur, and R. Schieder. 1987. "Unsharp Particle-Wave Duality in a Photon Split-Beam Experiment." *Foundations of Physics* 17: 891–903.
Moreva, Ekaterina, Giorgio Brida, Marco Gramegna, Vittorio Giovannetti, Lorenzo Maccone, and Marco Genovese. 2014. "Time From Quantum Entanglement: An Experimental Illustration." In *Physical Review A* 89, 052122. Arxiv.org/abs/1310.4691.
Mossbridge, Julia, Patrizio Tressoldi, and Jessica Utts. 2012. "Predictive Physiological Anticipation Preceding Seemingly Unpredictable Stimuli: A Meta-Analysis." *Frontiers in Psychology* 3: article 390. https://doi.org/10.3389/fpsyg.2012.00390.
Nash, John. 1951. "Non-Cooperative Games." *The Annals of Mathematics* 54: 286–295.
Neumann, Johann von. (1932) 1996. *Mathematische Grundlagen der Quantenmechanik*. 2nd ed. Berlin: Springer-Verlag.
Neumann, Johann von, and Oskar Morgenstern. (1947) 1953. *Theory of Games and Economic Behavior*. Princeton, NJ: Princeton University Press.
Nichols, Shaun, and Joshua Knobe. 2007. "Moral Responsibility and Determinism: The Cognitive Science of Folk Intuitions." *Noûs* 41: 663–685.
Nichols, Shaun. 2011. "Experimental Philosophy and the Problem of Free Will." *Science* 331: 1401–1403.
O'Connell, Aaron D., Max Hofheinz, Markus Ansmann, Radoslaw C. Bialczak, Mike Lenander, Erik Lucero, Matthew Neeley, et al. 2010. "Quantum Ground State and Single-Phonon Control of a Mechanical Resonator." *Nature* 464: 697–703.

Page, Don N. 1995. "Sensible Quantum Mechanics: Are Probabilities Only in the Mind?" University of Alberta preprint. Accessed October 2015. arXiv:gr-qc/9507024v1.
Page, Don N. 2011. "Consciousness and the Quantum." University of Alberta preprint. Accessed January 2016. arXiv:1102.5339v1.
Page, Don N., and William K. Wootters. 1983. "Evolution without Evolution: Dynamics Described by Stationary Observables." *Physical Review* 27: 2885–2892.
Patel, Vihan M., and Charles H. Lineweaver. 2015. *An Inflationary Explanation to the Initial Entropy Problem*. Second International Electronic Conference on Entropy and its Applications. November 15–30, 2015. Conference Proceedings Paper – Entropy. https://www.sciforum.net/conference/ecea-2. Accessed August 28, 2018.
Penrose, Roger. 1994. *Shadows of the Mind: A Search for the Missing Science of Consciousness*. Oxford: Oxford University Press.
Petkov, Vesselin. 2005. *Relativity and the Nature of Spacetime*. Berlin: Springer-Verlag.
Plato. (~ 360 B.C.) 2000. *Timaeus*. Translated by Donald J. Zeyl. Indianapolis, IN: Hackett Publishing Company, Inc.
Polkinhorne, John. C. 1984. *The Quantum World*. London: Longman.
Popper, Karl R. (1956) 1999. *Realism and the Aim of Science*. 2nd ed. Guildford and King's Lynn (UK): Biddles.
Popper, Karl R. 1959. *The Logic of Scientific Discovery*. Oxford: Routledge.
Popper, Karl R. 2014. *Conjectures and Refutations: The Growth of Scientific Knowledge*. Oxford: Routledge.
Price, Huw, and Ken Wharton. 2015. "Disentangling the Quantum World." *Entropy* 17: 7752–7767.
Racker, Heinrich. 2001. *Transference and Counter-Transference*. Madison, CT: International Universities Press.
Radin, Dean, Leena Michel, Karla Galdamez, Paul Wendland, Robert Rickenbach, and Amaud Delorme. 2012. "Consciousness and the Double-Slit Interference Pattern: Six Experiments." *Physics Essays* 25: 157–171.
Rapoport, Amnon. 1995. "Individual Strategies in a Market Entry Game." *Group Decision and Negotiation* 4: 117–133.
Raspe, Rudolf E. 1785. *The Surprising Adventures of Baron Munchhausen*. London: Smith.
Read, Daniel. 2007. "Experienced Utility: Utility Theory from Jeremy Bentham to Daniel Kahneman." *Thinking & Reasoning* 13: 45–61.
Redelmeier, Donald. A., Joel Katz, and Daniel Kahneman. 2003. "Memories of a Colonoscopy: A Randomized Trial." *Pain* 104: 187–194.
Riedel, C. Jess. 2013. "Is the Preferred Basis Problem Solved?" Answer to a question in https://www.physics.stackexchange.com/questions/65177. Accessed September 18, 2015.
Rilling, James K., Alan G. Sanfey, Jessica A. Aronson, Leigh E. Nystrom, and Jonathan D. Cohena. 2004. "The Neural Correlates of Theory of Mind within Interpersonal Interactions." *Neuroimage* 22: 1694–1703.
Rist, Ray. 1970. "Student Social Class and Teacher Expectations: The Self-Fulfilling Prophecy in Ghetto Education." *Harvard Educational Review* 40: 411–451.
Rosenthal, Robert. 1956. "An Attempt at the Experimental Induction of the Defense Mechanism of Projection." Unpublished doctoral dissertation, University of California at Los Angeles.
Rosenthal, Robert. 1976. *Experimenter Effects in Behavioral Research*. Enlarged edition. New York et al.: Irvington Publishers.
Rosenthal, Robert, and Kermit L. Fode. 1963. "The Effect of Experimenter Bias on the Performance of the Albino Rat." *Behavioral Science* 8: 183–189.
Rosenthal, Robert, and Lenore Jacobson. 1968. "Pygmalion in the Classroom." *The Urban Review* 3: 16–20.
Rosenthal, Robert, and Donald B. Rubin. 1978. "Interpersonal Expectancy Effects: The First 345 Studies." *Behavioral and Brain Sciences* 1: 377–415.

Samuelson, William, and Richard Zeckhauser. 1988. “Status Quo Bias in Decision Making.” *Journal of Risk and Uncertainty* 1: 7–59.

Sandri, Serena, Christian D. Schade, Oliver Mußhoff, and Marting Odening. 2010. “Holding On for Too Long? - An Experimental Study on Inertia in Entrepreneurs' and Non-Entrepreneurs' Disinvestment Choices.” *Journal of Economic Behavior and Organization* 76: 30–44.

Sarasvathy, Saras. 2001. “Causation and Effectuation: Toward a Theoretical Shift from Economic Inevitability to Entrepreneurial Contingency.” *Academy of Management Review* 26: 243–263.

Saunders, Simon. 1993. “Decoherence, Relative States, and Evolutionary Adaption.” *Foundations of Physics* 23: 1553–1585.

Saunders, Simon. 1995. “Time, Quantum Mechanics, and Decoherence.” *Synthese* 102: 235–266.

Saunders, Simon, Jonathan Barrett, Adrian Kent, and David Wallace, eds. 2012. *Many Worlds? Everett, Quantum Theory, & Reality.* Oxford: Oxford University Press.

Savage, Leonard J. 1954. *The Foundations of Statistics.* New York: John Wiley.

Schade, Christian D. 2015. “Collecting Evidence for the Permanent Coexistence of Parallel Realities: An Interdisciplinary Approach.” *Journal of Cognition and Neuroethics* 3 (1): 327–362.

Schade, Christian D., Howard C. Kunreuther, and Philipp Koellinger. 2012. “Protecting Against Low-Probability Disasters: The Role of Worry.” *Journal of Behavioral Decision Making* 25: 534–543.

Schade-Strohm, Tanja. 2017. *Werde zur besten Version Deines Selbst – Aus der Verstimmung in die Bestimmung*. Bielefeld: TAO. [Becoming the Best Version of Yourself: In Tune with Your Destiny].

Schelling, Thomas C. 1978. “Egonomics, or the Art of Self-Management.” *American Economic Review* 68: 290–294.

Schelling, Thomas C. (1960) 1981. *The Strategy of Conflict.* Boston: Harvard University Press.

Schooler, J. W. 2014. “Metascience Could Rescue the 'Replication Crisis.'” *Nature* 515 (7525): 9.

Schopenhauer, Arthur. (1818) 2010. “The World as Will and Representation.” In *The World as Will and Representation*, edited by Christopher Janaway, vol. 1, 1st ed. Cambridge: Cambridge University Press.

Schrödinger, Erwin. 1926. “Quantisierung als Eigenwertproblem.” *Annalen der Physik* 79: 361–376.

Schrödinger, Erwin. (1935) 1983. “The Present Situation in Quantum Mechanics.“ Reprinted in *Quantum Theory and Measurement*, edited by John A. Wheeler and Wojciech H. Zurek, 152–167. Princeton, NJ: Princeton University Press. (Original German title: Die gegenwärtige Situation in der Quantenmechanik.)

Schrödinger, Erwin. (1952) 1995. “July Colloquium, 1952.” In *Erwin Schrödinger: The Interpretation of Quantum Mechanics: Dublin Seminars (1949–1955) and Other Unpublished Essays*, edited and with introduction by Michael Bitbol, 19–37. Woodbridge, CT: Ox Bow Press.

Schrödinger, Erwin. (1958) 2004. *Mind and Matter.* Jointly reprinted edition with *What Is Life?: The Physical Aspects of the Living Cell* and *Autobiographical Sketches.* Cambridge et al.: Cambridge Univ. Press.

Schultze-Kraft, Matthias, Daniel Birnam, Marco Rusconi, Carsten Allefeld, Kai Görgen, Sven Dähne, Benjamin Blanker, and John-Dylan Hayes. 2015. “The Point of Non Return in Vetoing Self-Initiated Movements.” PNAS Early Edition. Accessed January 3, 2016. http://www.pnas.org/content/113/4/1080.

Science Daily. 2014. “Discovery of Quantum Vibrations in ‘Microtubules’ inside Brain Neurons Supports Controversial Theory of Consciousness.” Date of publication January 16, 2014. Accessed April 12, 2014. http://www.sciencedaily.com/releases/2014/01/140116085105.htm.

Scully, Marian O., Berthold-Georg Englert, and Herbert Walther. 1991. “Quantum Optical Tests of Complementarity.” *Nature* 351: 111–116.

Selten, Reinhard, and Werner Güth. 1982. "Equilibrium Point Selection in a Class of Market Entry Games." In *Games, Economic Dynamics, Time Series Analysis: A Symposium in Memoriam Oscar Morgenstern*, edited by M. Deistler, E. Fürst, and G. Schwödiauer, 101–116. Wien: Physica-Verlag.

Sheehan, Daniel P., ed. 2006. *Frontiers of Time: Retrocausation – Experiment and Theory*. Melville, NY: American Institute of Physics.

Silberstein, Michael, W. M. Stuckey, and Timothy McDevitt. 2018. *Beyond the Dynamical Universe: Unifying Block Universe Physics and Time as Experienced.* Oxford: Oxford University Press.

Simon, Charles W., and William H. Evans. 1956. "EEG, Consciousness and Sleep." *Science* 124: 1066.

Simon, Herbert. 1955. "A Behavioral Model of Rational Choice." *Quarterly Journal of Economics* 69: 99–118.

Simon, Herbert. 1956. "Rational Choice and the Structure of the Environment." *Psychological Review* 63: 129–138.

Simon, Herbert. 1957. *Models of Man: Social and Rational*. New York: John Wiley and Sons.

Simonton, O. Carl, and Stephanie S. Simonton. 1975. "Belief Systems and Management of the Emotional Aspects of Malignancy." *Journal of Transpersonal Psychology* 1: 29–47.

Skow, Bradford. 2009. "Relativity and the Moving Spotlight." *The Journal of Philosophy* 106: 666–678.

Skow, Bradford. 2012. "Why Does Time Pass?" *Noûs* 46: 223–242.

Soon, Chun S., Marcel Brass, Hans-Jochen Heinze, and John-Dylan Haynes. 2008. "Unconscious Determinants of Free Decisions in the Human Brain." *Nature Neuroscience* 11: 543–554.

Squires, Euan J. 1988. "The Unique World of the Everett Version of Quantum Theory." *Foundations of Physics Letters* 1: 13–20.

Squires, Euan J. 1991. "One Mind or Many – a Note on the Everett Interpretation of Quantum Theory." *Synthese* 89: 283–286.

Squires, Euan J. 1992. "History and Many-Worlds Quantum Theory." *Foundations of Physics Letters* 5: 279–290.

Squires, Euan J. 1994. *The Mystery of the Quantum World.* 2nd ed. New York, Milton Park: Taylor & Francis Group.

Stapp, Henry P. 2009. *Mind, Matter, and Quantum Mechanics*. 3rd ed. Berlin: Springer-Verlag.

Stefana, Alberto. 2017. *History of Countertransference. From Freud to the British Object Relations School*. London and New York: Routledge.

Strawson, Peter F. (1962) 1993. "Freedom and Resentment." *Proceedings of the British Academy* 48: 1–25. Reprinted in John Martin Fischer and Mark Ravizza, eds. *Perspectives on Moral Responsibility*. Ithaca, NY: Cornell University Press.

Taleb, Nassim N. 2007. *The Black Swan. The Impact of the Highly Improbable*. New York: Random House.

Tegmark, Max. 2004. "Parallel Universes." In *Science and Ultimate Reality: From Quantum to Cosmos, Honoring John Wheeler's 90th Birthday,* edited by John D. Barrow, Paul C.W. Davies, and Charles L. Harper, 459–491. Cambridge et al: Cambridge University Press.

Thaler, Richard, and Eric J. Johnson. 1990. "Gambling with the House Money and Trying to Break Even: The Effects of Prior Outcomes on Risky Choice." *Management Science* 36: 643–660.

Thomas, William I., and Dorothy Swaine Thomas. 1928. *The Child in America: Behavior Problems and Programs*. New York: Knopf.

Tietz, Rainer. 1988. "Experimental Economics: Ways to Model Bounded Rational Bargaining Behavior." *Lecture Notes in Economics and Mathematical Systems* 314. New York et al.: Springer.

Triantaphyllou, Evangelos. 2000. *Multi-Criteria Decision Making Methods: A Comparative Study (Applied Optimization).* Dordrecht (NL): Kluwer Academic Publishers.

Tversky, Amos, and Daniel Kahneman. 1974. "Judgment under Uncertainty: Heuristics and Biases." *Science* 185: 1124–1131.

Tversky, Amos, and Daniel Kahneman. 1991. "Loss Aversion in Riskless Choice: A Reference-Dependent Model." *The Quarterly Journal of Economics* 106: 1039–1061.

Tversky, Amos, and Daniel Kahneman. 1992. "Advances in Prospect Theory: Cumulative Representation of Uncertainty." *Journal of Risk and Uncertainty* 5: 297–323.

Tversky, Amos, and Itamar Simonson. 1993. "Context-Dependent Preferences." *Management Science* 39: 1179–1189.

Vaidman, Lev. 1998. "Time-Symmetrized Quantum Theory." *Fortschritte der Physik* 46: 729–739.

Vaidman, Lev. 1994. "On the Paradoxical Aspects of New Quantum Experiments." *PSA: Proceedings of the Biennial Meeting of the Philosophy of Science Association* 1: 211–217.

Vaidman, Lev. 2012. "Time-Symmetry and the Many-Worlds Interpretation." In *Many Worlds? Everett, Quantum Theory, & Reality*, edited by Simon Saunders, Jonathan Barrett, Adrian Kent, and David Wallace, 582–596. Oxford: Oxford University Press.

Valdano, Domino. 2014. "How Does The Many Worlds Interpretation Explain (Away) Nonlocality?" Online contribution to Quora: http://www.quora.com/how-does-the-many-worlds-interpretation-explain-away-nonlocality. Answered April 23, 2014. Accessed March 29, 2018.

Van Daal, Jan. 1996. "From Utilitarianism to Hedonism: Gossen, Jevons and Walras." *Journal of the History of Economic Thought* 18: 271–286.

Van der Kolk, Bessel A., and Martin S. Greenberg. 1987. "The Psychobiology of the Trauma Response: Hyperarousal, Constriction, and Addiction to Traumatic Reexposure." In *Psychological Trauma*, edited by Bessel A. van der Kolk, 63–87. Washington, D.C.: American Psychiatric Press.

Van Inwagen, Peter. 1994. "When the Will Is Not Free." *Philosophical Studies* 75: 95–113.

Velmans, Max. 2003. "Preconscious Free Will." *Journal of Consciousness Studies* 10: 42–61.

Vollmer, Gerhard. 1999. "Die vierte bis siebte Kränkung des Menschen." In *Arbeitsgruppe Mensch – Technik – Umwelt*, edited by Hans-Hermann Franzke, Technische Universität Berlin. *Schriftenreihe Technik und Gesellschaft* 3: 67–85.

Wallace, David. 2002. "Worlds in the Everett Interpretation." *Studies in the History and Philosophy of Modern Physics* 33: 637–661.

Wallace, David. 2012a. "Decoherence and Ontology (or: How I learned to Stop Worrying and Love FAPP)." In *Many Worlds? Everett, Quantum Theory, & Reality*, edited by Simon Saunders, Jonathan Barrett, Adrian Kent, and David Wallace, 53–72. Oxford: Oxford University Press.

Wallace, David. 2012b. "How to Prove the Born Rule." In *Many Worlds? Everett, Quantum Theory, & Reality*, edited by Simon Saunders, Jonathan Barrett, Adrian Kent, and David Wallace, 227–263. Oxford: Oxford University Press.

Wallace, David. 2012c. *The Emergent Multiverse: Quantum Theory According to the Everett Interpretation*. Oxford: Oxford University Press.

Walter, Henrik. 2001. *Neurophilosophy of Free Will. From Libertarian Illusions to a Concept of Natural Autonomy*, translated by Cynthia Klohr. Cambridge: A Bradford Book – The MIT Press.

Watzlawick, Paul. 1976. *How Real Is Real?* New York: Random House.

Watzlawick, Paul. 1978. *The Language of Change: Elements of Therapeutic Communication*. New York: Basic Books.

Watzlawick, Paul. 1984a. "Self-Fulfilling Prophecies." In *The Production of Reality: Essays and Readings on Social Interaction*, 5th ed., edited by Jodi O'Brian, 382–394. Thousand Oaks, CA: Pine Forge Press.

Watzlawick, Paul. ed. 1984b. *The Invented Reality: How Do We Know What We Believe We Know? (Contributions to Constructivism)*. New York: Norton.

Whyte, William F. 1943. *Street Corner Society*. Chicago, IL: University of Chicago Press.

Wigner, Eugene P. (1961) 1983. "Remarks on the Mind-Body Question." Reprinted in *Quantum Theory and Measurement*, edited by John A. Wheeler and Wojciech H. Zurek, 285-302. Princeton, NJ: Princeton University Press. (Original German title: Die gegenwärtige Situation in der Quantenmechanik.)

Woodfield, Andrew. (1976) 2010. *Teleology*. 2nd ed. Cambridge: Cambridge University Press.

Wundt, Wilhelm. 1894. "Über psychische Kausalität und das Prinzip des psychophysischen Parallelismus." In *Philosophische Studien* 10: 1–124. (English translation of the title: About psychological causality and the principle of psychophysical parallelism.)

Yerkes, Robert M., and John D. Dodson. 1908. "The Relation of Strength of Stimulus to Rapidity of Habit-Formation." *Journal of Comparative Neurology and Psychology* 18: 459–482.

Yu, Shan, and Danko Nikolic. 2011. "Quantum Mechanics Needs No Consciousness." *Annalen der Physik* 523: 931–938.

Zeh, H. Dieter. 1970. "On the Interpretation of Measurement in Quantum Theory." *Foundations of Physics* 1: 69–76.

Zeh, H. Dieter. 1999. *The Physical Basis of the Direction of Time*. 3rd ed. Berlin: Springer-Verlag.

Zeh, H. Dieter. 2012. *Physik ohne Realität: Tiefsinn oder Wahnsinn?* Berlin – Heidelberg: Springer-Verlag.

Zeh, H. Dieter. 2013. "The Role of the Observer in the Everett Interpretation." *NeuroQuantology* 11: 97–105.

Zeh, H. Dieter. 2016. "John Bell's Varying Interpretations of Quantum Mechanics – Memories and Comments." In *Quantum Nonlocality and Reality*, edited by Mary Bell and Shan Gao. Cambridge: Cambridge University Press. arxiv:1402.5498v8.

Zeh, H. Dieter. 2016a. "The Strange (Hi)story of Particles and Waves." *Zeitschrift für Naturforschung A* 71: 195–212.

Zeilinger, Anton. 1999. "Experiment and the Foundations of Quantum Physics." *Review of Modern Physics* 71: S288-S297.

Zuckerman, Marvin. 1979. *Sensation Seeking: Beyond the Optimal Level of Arousal*. New York: Wiley.

Zurek, Wojciech H. 1991. "Decoherence and the Transition from Quantum to Classical." *Physics Today* 44: 36–44.

Zurek, Wojciech H. 2002. "Decoherence and the Transition from Quantum to Classical – Revisited." *Los Alamos Science* 27: 2–25.

Printed by Printforce, the Netherlands